THOMAS TURNER TATE

THE FIRST THREE BOOKS OF EUCLID'S ELEMENTS OF GEOMETRY

FROM THE

TEXT OF DR. ROBERT SIMSON

TOGETHER WITH

VARIOUS USEFUL THEOREMS AND PROBLEMS

AS GEOMETRICAL EXERCISES ON EACH BOOK

Elibron Classics
www.elibron.com

Elibron Classics series.

ISBN 1-4021-5882-3 (paperback)
ISBN 1-4021-2205-5 (hardcover)

This Elibron Classics Replica Edition is an unabridged facsimile of the edition published in 1866 by Longmans, Green, and Co., London.

THE

FIRST THREE BOOKS

OF

EUCLID'S ELEMENTS

OF

GEOMETRY

FROM THE

TEXT OF DR. ROBERT SIMSON

TOGETHER WITH

VARIOUS USEFUL THEOREMS AND PROBLEMS

AS GEOMETRICAL EXERCISES ON EACH BOOK

BY THOMAS TATE

OF KNELLER TRAINING COLLEGE, TWICKENHAM

AUTHOR OF

'THE PRINCIPLES OF THE DIFFERENTIAL AND INTEGRAL CALCULUS'

ETC.

NEW EDITION

LONDON

LONGMANS, GREEN, AND CO.

1866

LONDON
PRINTED BY SPOTTISWOODE AND CO.
NEW-STREET SQUARE

ADVERTISEMENT.

THIS edition of Euclid has been published in a cheap form, with the hope that it may tend to advance the mathematical education of this country, and with an especial reference to the instruction of Schoolmasters who may be desirous of obtaining a Government certificate.

EUCLID'S

ELEMENTS OF GEOMETRY.

BOOK I.

DEFINITIONS.

I.

A POINT is that which hath no parts, or which hath no magnitude.

II.

A line is length without breadth.

III.

The extremities of a line are points.

IV.

A straight line is that which lies evenly between its extreme points.

V.

A superficies is that which hath only length and breadth.

VI.

The extremities of a superficies are lines.

VII.

A plane superficies is that in which any two points being taken, the straight line between them lies wholly in that superficies.

VIII.

" A plane angle is the inclination of two lines to one another in a plane, which meet together, but are not in the same direction."

IX.

A plane rectilineal angle is the inclination of two straight lines to one another, which meet together, but are not in the same straight line.

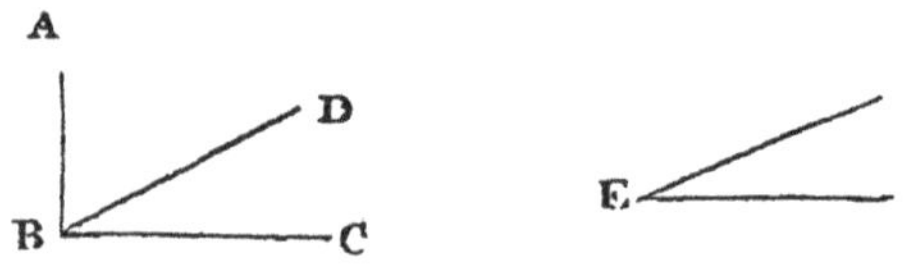

N.B. "When several angles are at one point B, any one of them is expressed by three letters, of which the letter that is at the vertex of the angle, that is, at the point in which the straight lines that contain the angle meet one another, is put between the other two letters, and one of these two is somewhere upon one of those straight lines, and the other upon the other line: Thus the angle which is contained by the straight lines AB, CB, is named the angle ABC, or CBA; that which is contained by AB, BD, is named the angle ABD, or DBA; and that which is contained by DB, CB is called the angle DBC, or CBD; but, if there be only one angle at a point, it may be expressed by a letter placed at that point; as the angle at E."

X.

When a straight line standing on another straight line makes the adjacent angles equal to one another, each of the angles is called a right angle; and the straight line which stands on the other is called a perpendicular to it.

XI.

An obtuse angle is that which is greater than a right angle.

XII.

An acute angle is that which is less than a right angle.

XIII.

"A term or boundary is the extremity of any thing."

XIV.

A figure is that which is inclosed by one or more boundaries.

XV.

A circle is a plane figure contained by one line, which is called the circumference, and is such that all straight lines drawn from a certain point within the figure to the circumference are equal to one another.

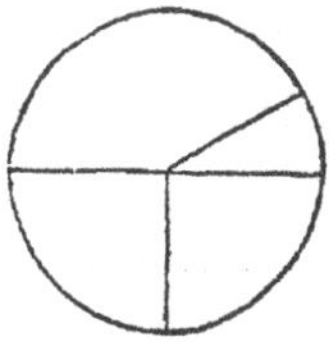

XVI.

And this point is called the centre of the circle.

XVII.

A diameter of a circle is a straight line drawn through the centre, and terminated both ways by the circumference.

XVIII.

A semicircle is the figure contained by a diameter and the part of the circumference cut off by the diameter.

XIX.

"A segment of a circle is the figure contained by a straight line, and the circumference it cuts off."

XX.

Rectilineal figures are those which are contained by straight lines.

XXI.

Trilateral figures, or triangles, by three straight lines.

XXII.

Quadrilateral, by four straight lines.

XXIII.

Multilateral figures, or polygons, by more than four straight lines.

XXIV.

Of three-sided figures, an equilateral triangle is that which has three equal sides.

XXV.

An isosceles triangle is that which has only two sides equal.

XXVI.

A scalene triangle, is that which has three unequal sides.

XXVII.

A right angled triangle, is that which has a right angle.

XXVIII.

An obtuse angled triangle, is that which has an obtuse angle.

XXIX.

An acute angled triangle, is that which has three acute angles.

XXX.

Of four-sided figures, a square is that which has all its sides equal, and all its angles right angles.

XXXI.

An oblong, is that which has all its angles right angles, but has not all its sides equal.

XXXII.

A rhombus, is that which has all its sides equal, but its angles are not right angles.

XXXIII.

A rhomboid, is that which has its opposite sides equal to one another, but all its sides are not equal, nor its angles right angles.

XXXIV.

All other four-sided figures besides these, are called Trapeziums.

XXXV.

Parallel straight lines are such as are in the same plane, and which, being produced ever so far both ways, do not meet.

POSTULATES.

I.

Let it be granted that a straight line may be drawn from any one point to any other point.

II.

That a terminated straight line may be produced to any length in a straight line.

III.

And that a circle may be described from any centre, at any distance from that centre.

AXIOMS.

I.

Things which are equal to the same are equal to one another.

II.

If equals be added to equals, the wholes are equal.

III.

If equals be taken from equals, the remainders are equal.

IV.

If equals be added to unequals, the wholes are unequal.

V.

If equals be taken from unequals, the remainders are unequal

VI.

Things which are double of the same, are equal to one another.

VII.

Things which are halves of the same, are equal to one another.

VIII.

Magnitudes which coincide with one another, that is, which exactly fill the same space, are equal to one another.

IX.

The whole is greater than its part.

X.

Two straight lines cannot inclose a space.

XI.

All right angles are equal to one another.

XII.

"If a straight line meets two straight lines, so as to make the two interior angles on the same side of it taken together less than two right angles, these straight lines being continually produced, shall at length meet upon that side on which are the angles which are less than two right angles."

Proposition I. Problem.

To describe an equilateral triangle upon a given finite straight line.

Let AB be the given straight line; it is required to describe an equilateral triangle upon it.

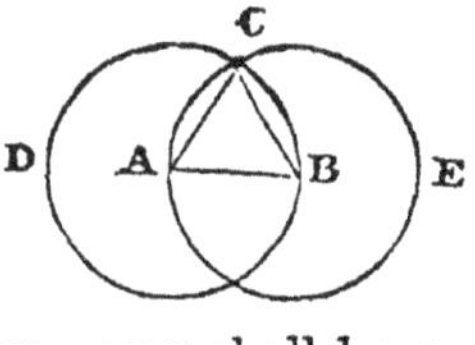

From the centre A, at the distance AB, describe (Postulate 3.) the circle BCD, and from the centre B, at the distance BA, describe the circle ACE; and from the point C, in which the circles cut one another, draw the straight lines (Post. 1.) CA, CB, to the points A, B; ABC shall be an equilateral triangle.

Because the point A is the centre of the circle BCD, AC is equal (Definition 15.) to AB; and because the point B is the centre of the circle ACE, BC is equal to BA: But it has been proved that CA is equal to AB; therefore CA, CB are each of them equal to AB; but things which are equal to the same are equal to one another (Axiom 1.); therefore CA is equal to CB; wherefore CA, AB, BC are equal to one another; and the triangle ABC is therefore equilateral, and it is described upon the given straight line AB. Which was required to be done.

PROP. II. PROB.

From a given point to draw a straight line equal to a given straight line.

Let A be the given point, and BC the given straight line; it is required to draw from the point A a straight line equal to BC.

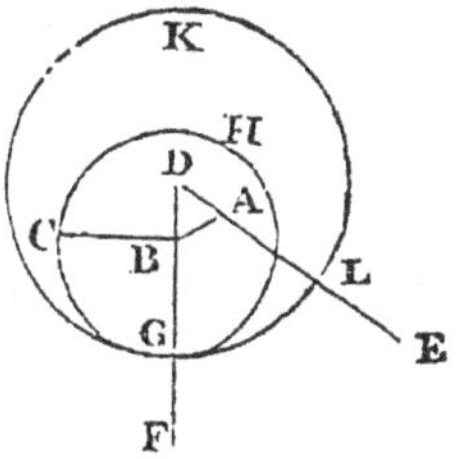

From the point A to B draw (Post. 1.) the straight line AB; and upon it describe (I. 1.) the equilateral triangle DAB, and produce (Post. 2.) the straight lines DA, DB, to E and F; from the centre B, at the distance BC, describe (Post. 3.) the circle CGH, and from the centre D, at the distance DG, describe the circle GKL. AL shall be equal to BC.

Because the point B is the centre of the circle CGH, BC is equal (Def. 15.) to BG; and because D is the centre of the circle GLK, DL is equal to DG, and DA, DB, parts of them, are equal (Constr.); therefore the remainder AL is equal to the remainder BG (Ax. 3.): But it has been shown that BC is equal to BG; wherefore AL and BC are each of them equal to BG; and things that are equal to the same are equal to one another; therefore the straight line AL is equal to BC. Wherefore from the given point A a straight line AL has been drawn equal to the given straight line BC. Which was to be done.

PROP. III. PROB.

From the greater of two given straight lines to cut off a part equal to the less.

Let AB and C be the two given straight lines, whereof AB is the greater. It is required to cut off from AB, the greater, a part equal to C, the less.

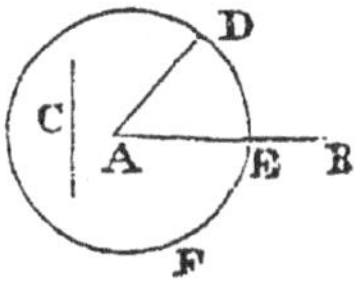

From the point A draw (I. 2.) the straight line AD equal to C; and from the centre A, and at the distance AD, describe (Post. 3.) the circle DEF: AE shall be equal to C.

Because A is the centre of the circle DEF, AE is equal to AD (Def. 15.); but the straight line C is likewise equal to AD (Constr.); whence AE and C are each of them equal to AD; wherefore the straight line AE is equal to C (Ax. 1.), and from AB, the greater of two straight lines, a part AE has been cut off equal to C the less. Which was to be done.

Prop. IV. Theorem.

If two triangles have two sides of the one equal to two sides of the other, each to each; and have likewise the angles contained by those sides equal to one another; they shall likewise have their bases, or third sides, equal; and the two triangles shall be equal; and their other angles shall be equal, each to each, viz. those to which the equal sides are opposite.

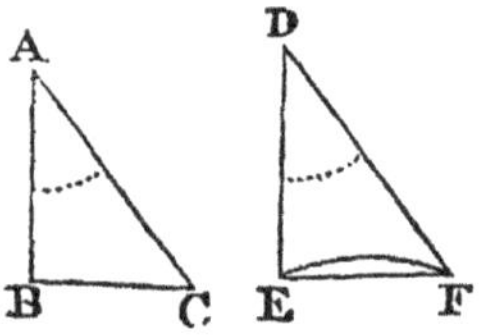

Let ABC, DEF be two triangles which have the two sides AB, AC equal to the two sides DE, DF, each to each, viz. AB to DE, and AC to DF; and the angle BAC equal to the angle EDF, the base BC shall be equal to the base EF; and the triangle ABC to the triangle DEF; and the other angles, to which the equal sides are opposite, shall be equal each to each, viz. the angle ABC to the angle DEF, and the angle ACB to DFE.

For, if the triangle ABC be applied to DEF, so that the point A may be on D, and the straight line AB upon DE; the point B shall coincide with the point E, because AB is equal to DE (Hyp.); and AB coinciding with DE, AC shall coincide with DF because the angle BAC is equal to the angle EDF (Hyp.); wherefore also the point C shall coincide with the point F, because the straight line AC is equal to DF (Hyp.). But the point B coincides with the point E; wherefore the base BC shall coincide with the base EF, because the point B coinciding with E, and C with F, if the base BC does not coincide with the base EF, two straight lines would inclose a space, which is impossible (Ax. 10.). Therefore the base BC shall coincide with the base EF, and be equal to it (Ax. 8.). Wherefore the whole triangle ABC shall coincide with the whole triangle DEF, and be equal to it; and the other angles of the one shall coincide with the remaining angles of the other, and be equal to them, viz. the angle ABC to the angle DEF, and the angle ACB to DFE. Therefore, if two triangles have two sides of the one equal to two sides of the other, each to each, and have likewise the angles contained by those sides equal to one another, their bases shall likewise be equal, and the triangles be equal, and their other angles to which the equal sides are opposite shall be equal, each to each. Which was to be demonstrated.

Prop. V. Theor.

The angles at the base of an Isosceles triangle are equal to one another; and, if the equal sides be produced, the angles upon the other side of the base shall be equal.

Let ABC be an Isosceles triangle, of which the side AB is equal to AC, and let the straight lines AB, AC be produced to D and E, the angle ABC shall be equal to the angle ACB, and the angle CBD to the angle BCE.

In BD take any point F, and from AE, the greater, cut off AG equal (I. 3.) to AF, the less, and join FC, GB.

Because AF is equal to AG, and AB to AC (Hyp.), the two sides FA, AC are equal to the two GA, AB, each to each; and they contain the angle FAG common to the two triangles AFC, AGB; therefore the base FC is equal (I. 4.) to the base GB, and the triangle AFC to the triangle AGB; and the remaining angles of the one are equal (I. 4.) to the remaining angles of the other, each to each, to which the equal sides are opposite; viz. the angle ACF to the angle ABG, and the angle AFC to the angle AGB: And because the whole AF is equal to the whole AG, of which the parts AB, AC, are equal: the remainder BF shall be equal (Ax. 3.) to the remainder CG; and FC was proved to be equal to GB; therefore the two sides BF, FC are equal to the two CG, GB, each to each; and the angle BFC is equal to the angle CGB, and the base BC is common to the two angles BFC, CGB; wherefore the triangles are equal (I. 4.), and their remaining angles, each to each, to which the equal sides are opposite; therefore the angle FBC is equal to the angle GCB, and the angle BCF to the angle CBG: And, since it has been demonstrated, that the whole angle ABG is equal to the whole ACF, the parts of which, the angles CBG, BCF are also equal; the remaining angle ABC is therefore equal to the remaining angle ACB, which are the angles at the base of the triangle ABC: And it has also been proved that the angle FBC is equal to the angle GCB, which are the angles upon the other side of the base. Therefore the angles at the base, &c. Q. E. D

COROLLARY. Hence every equilateral triangle is also equiangular.

PROP. VI. THEOR.

If two angles of a triangle be equal to one another, the sides also which subtend, or are opposite to, *the equal angles, shall be equal to one another.*

Let ABC be a triangle having the angle ABC equal to the angle ACB; the side AB is also equal to the side AC.

For, if AB be not equal to AC, one of them is greater than the other: let AB be the greater, and from it cut (I. 3.) off DB equal to AC, the less, and join DC; therefore, because in the triangles DBC, ACB, DB is equal to AC, and BC common to both, the two sides DB, BC are equal to the two AC, CB, each to each; and the angle DBC is equal to the angle ACB (Hyp.); therefore the base DC is equal to the base AB, and the triangle DBC is equal to the triangle (I. 4.) ACB, the less to the greater; which is absurd. Therefore AB is not unequal to AC, that is, it is equal to it. Wherefore, if two angles, &c. Q. E. D.

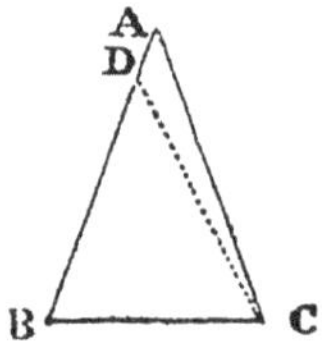

COR. Hence every equiangular triangle is also equilateral.

PROP. VII. THEOR.

Upon the same base, and on the same side of it, there cannot be two triangles that have their sides which are terminated in one extremity of the base equal to one another, and likewise those which are terminated in the other extremity.

If it be possible, let there be two triangles ACB, ADB, upon the same base AB, and upon the same side of it, which have their sides CA, DA, terminated in the extremity A of the base equal to one another, and likewise their sides CB, DB, and are terminated in B.

Join CD; then, in the case in which the vertex of each of the triangles is without the other triangle, because AC is equal to AD, the angle ACD is equal (I. 5.) to the angle ADC: But the angle ACD is greater than the angle BCD (Ax. 9.); therefore the angle ADC is greater also than BCD; much more then is the angle BDC greater than the angle BCD. Again, because CB is equal to DB, the angle

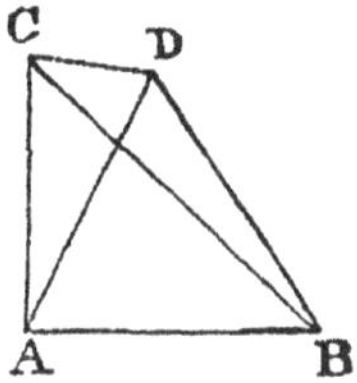

BDC is equal (I. 5.) to the angle BCD; but it has been demonstrated to be greater than it; which is impossible.

But if one of the vertices, as D, be within the other triangle ACB; produce AC, AD to E, F; therefore, because AC is equal to AD in the triangle ACD, the angles ECD, FDC upon the other side of the base CD are equal (I. 5.) to one another, but the angle ECD is greater than the angle BCD; wherefore the angle FDC is likewise greater than BCD; much more then is the angle BDC greater than the angle BCD. Again, because CB is equal to DB, the angle BDC is equal (I. 5.) to the angle BCD; but BDC has been proved to be greater than the same BCD; which is impossible. The case in which the vertex of the triangle is upon a side of the other, needs no demonstration.

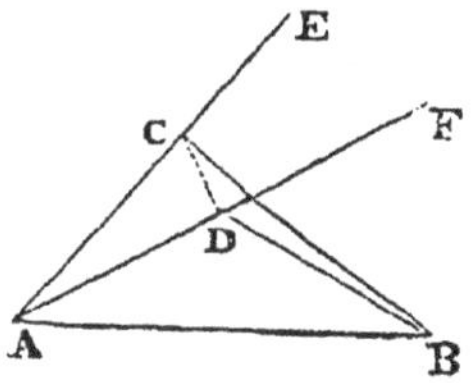

Therefore, upon the same base, and on the same sides of it, there cannot be two triangles, that have their sides which are terminated in one extremity of the base equal to one another, and likewise those which are terminated in the other extremity. Q. E. D.

Prop. VIII. Theor.

If two triangles have two sides of the one equal to two sides of the other, each to each, and have likewise their bases equal; the angle which is contained by the two sides of the one shall be equal to the angle contained by the sides equal to them, of the other.

Let ABC, DEF be two triangles having the two sides AB, AC, equal to the two sides, DE, DF, each to each, viz. AB to DE, and AC to DF; and also the base BC equal to the base EF. The angle BAC is equal to the angle EDF.

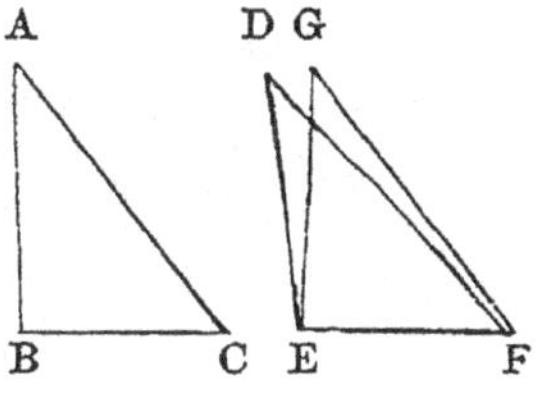

For if the triangle ABC be applied to DEF, so that the point B be on E, and the straight line BC upon EF; the point C shall also coincide with the point F, because BC is equal to EF (Hyp.); therefore BC coinciding with EF, BA and AC shall coincide with ED and DF; for, if the base BC coincides with the base EF, but the sides BA, CA do not coincide with the sides ED, FD, but have a different situation as EG, FG; then, upon the same base EF, and upon the same side of it, there can be two triangles that have their sides which are terminated in one extremity of the base equal to one another, and likewise their sides terminated in the other extremity: But this is impossible (I. 7.); therefore, if the base BC coincides with the base EF; the sides BA, AC cannot but coincide with the sides ED, DF; wherefore likewise the angle BAC coincides with the angle EDF, and is equal (Ax. 8.) to it. Therefore if two triangles, &c. Q. E. D.

PROP. IX. PROB.

To bisect a given rectilineal angle, that is, to divide it into two equal angles.

Let BAC be the given rectilineal angle, it is required to bisect it.

Take any point D in AB, and from AC cut (I. 3.) off AE equal to AD; join DE, and upon it, on the side towards B and C, describe (I. 1.) an equilateral triangle DEF; then join AF; the straight line AF bisects the triangle BAC.

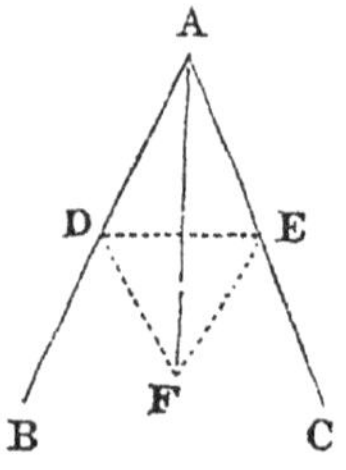

Because AD is equal to AE, and AF is common to the two triangles DAF, EAF; the two sides DA, AF, are equal to the two sides EA, AF, each to each; and the base DF is equal to the base EF; therefore the angle DAF is equal (I. 8.) to the angle EAF; wherefore the given rectilineal angle BAC is bisected by the straight line AF. Which was to be done.

PROP. X. PROB.

To bisect a given finite straight line, that is, to divide it into two equal parts.

Let AB be the given straight line; it is required to divide it into two equal parts.

Describe (I. 1.) upon it an equilateral triangle ABC, and bisect (I. 9.) the angle ACB by the straight line CD. AB is cut into two equal parts in the point D.

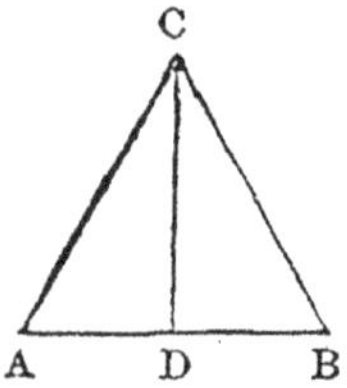

Because AC is equal to CB, and CD common to the two triangles ACD, BCD; the two sides AC, CD are equal to BC, CD, each to each; and the angle ACD is equal to the angle BCD (Constr.); therefore the base AD is equal to the base (I. 4.) DB, and the straight line AB is divided into two equal parts in the point D. Which was to be done.

Prop. XI. Prob.

To draw a straight line at right angles to a given straight line, from a given point in the same.

Let AB be a given straight line, and C a point given in it; it is required to draw a straight line from the point C at right angles to AB.

Take any point D in AD, and (I. 3.) make CE equal to CD, and upon DE describe (I. 1.) the equilateral triangle DFE, and join FC; the straight line FC drawn from the given point C is at right angles to the given straight line AB.

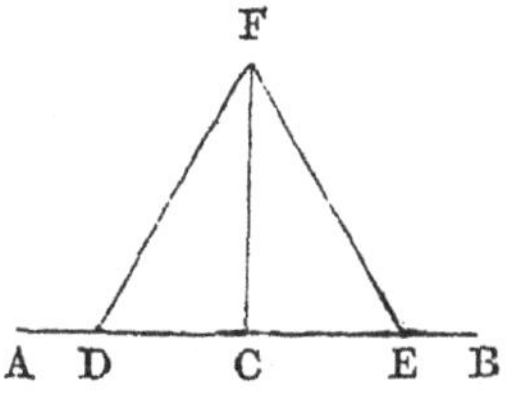

Because DC is equal to CE, and FC common to the two triangles DCF, ECF; the two sides DC, CF, are equal to the two EC, CF, each to each; and the base DF is equal to the base EF; therefore the angle DCF is equal (I. 8.) to the angle ECF; and they are adjacent angles: but, when the adjacent angles which one straight line makes with another straight line are equal to one another, each of them is called a right (I. Def. 10.) angle; therefore each of the angles DCF, ECF, is a right angle. Wherefore, from the given point C, in the given straight line AB, FC has been drawn at right angles to AB. Which was to be done.

Cor. By help of this problem, it may be demonstrated that two straight lines cannot have a common segment.

If it be possible, let the two straight lines ABC, ABD have the segment AB common to both of them. From the point B draw BE at right angles to AB; and because ABC is a straight line, the angle CBE is equal (I. Def. 10.) to the angle EBA; in the same manner, because ABD is a straight line, the angle DBE is equal to the angle EBA; wherefore the angle DBE is equal to the angle CBE, the less to the greater, which is impossible; therefore two straight lines cannot have a common segment.

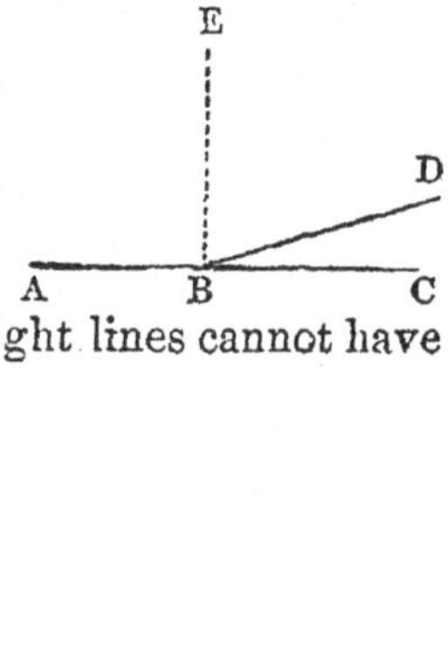

PROP. XII. PROB.

To draw a straight line perpendicular to a given straight line of an unlimited length, from a given point without it.

Let AB be the given straight line, which may be produced to any length both ways, and let C be a point without it. It is required to draw a straight line perpendicular to AB from the point C.

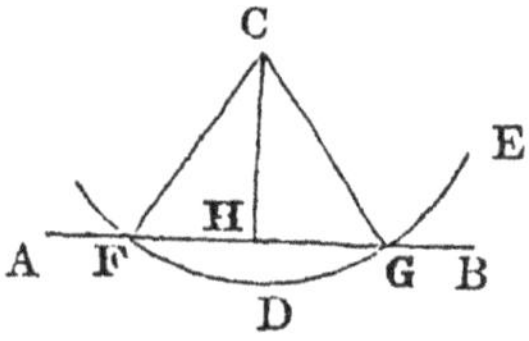

Take any point D upon the other side of AB, and from the centre C, at the distance CD, describe (Post. 3.) the circle EGF meeting AB in F, G; and bisect (I. 10.) FG in H, and join CF, CH, CG; the straight line CH, drawn from the given point C, is perpendicular to the given straight line AB.

Because FH is equal to HG, and HC common to the two triangles FHC, GHC, the two sides FH, HC are equal to the two GH, HC, each to each; and the base CF is equal (I. Def. 15.) to the base CG; therefore the angle CHF is equal (I. 8.) to the angle CHG; and they are adjacent angles; but when a straight line standing on a straight line makes the adjacent angles equal to one another, each of them is a right angle, and the straight line which stands upon the other is called a perpendicular to it (Def. 10.); therefore from the given point C a perpendicular CH has been drawn to the given straight line AB. Which was to be done.

PROP. XIII. THEOR.

The angles which one straight line makes with another upon the one side of it, are either two right angles, or are together equal to two right angles.

Let the straight line AB make with CD, upon one side of it, the angles CBA, ABD; these are either two right angles, or are together equal to two right angles.

For, if the angle CBA be equal to ABD, each of them is a right (Def. 10.) angle; but, if not, from the point B draw BE at right angles (I. 11.) to CD; therefore the angles CBE, EBD are two right angles (Def. 10.); and because CBE is equal to the two angles CBA, ABE together, add the angle EBD to each of these equals; therefore the angles CBE, EBD are equal (Ax.

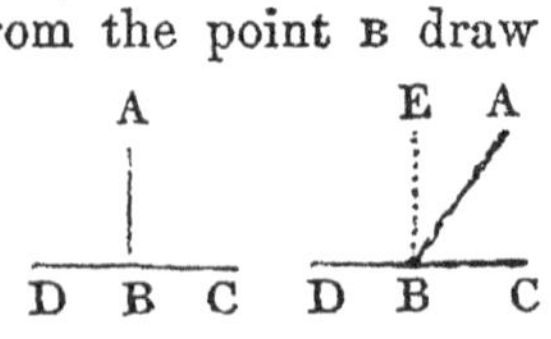

2.) to the three angles CBA, ABE, EBD. Again, because the angle DBA is equal to the two angles DBE, EBA, add to these equals the angle ABC; therefore the angles DBA, ABC are equal to the three angles DBE, EBA, ABC; but the angles CBE, EBD have been demonstrated to be equal to the same three angles; and things that are equal to the same are equal (Ax. 1.) to one another; therefore the angles CBE, EBD are equal to the angles DBA, ABC; but CBE, EBD are two right angles; therefore DBA, ABC are together equal to two right angles. Wherefore, when a straight line, &c. Q.E.D.

Prop. XIV. Theor.

If, at a point in a straight line, two other straight lines, upon the opposite sides of it, make the adjacent angles together equal to two right angles, these two straight lines shall be in one and the same straight line.

At the point B in the straight line AB, let the two straight lines BC, BD upon the opposite sides of AB, make the adjacent angles ABC, ABD, equal together to two right angles. BD is in the same straight line with CB.

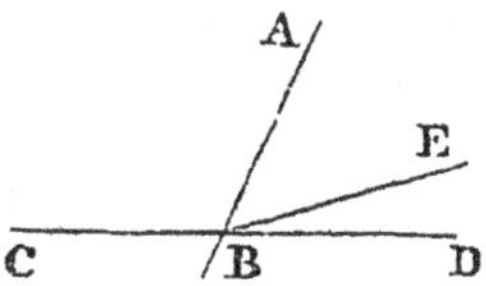

For, if BD be not in the same straight line with CB, let BE be in the same straight line with it; therefore, because the straight line AB makes angles with the straight line CBE, upon one side of it, the angles ABC, ABE are together equal (I. 13.) to two right angles; but the angles ABC, ABD are likewise together equal to two right angles (Hyp.); therefore the angles CBA, ABE are equal to the angles CBA, ABD. Take away the common angle ABC, the remaining angle ABE is equal (Ax. 3.) to the remaining angle ABD, the less to the greater, which is impossible; therefore BE is not in the same straight line with BC. And, in like manner, it may be demonstrated that no other can be in the same straight line with it but BD, which therefore is in the same straight line with CB. Wherefore, if at a point, &c. Q.E.D.

Prop. XV. Theor.

If two straight lines cut one another, the vertical, or opposite, angles shall be equal.

Let the two straight lines AB, CD cut one another in the

point E; the angle AEC shall be equal to the angle DEB, and CEB to EAD.

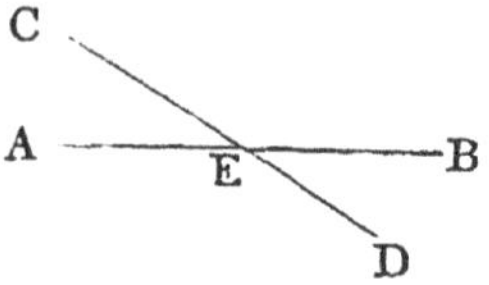

Because the straight line AE makes with CD the angles CEA, AED, these angles are together equal (I. 13.) to two right angles. Again, because the straight line DE makes with AB the angles AED, DEB, these also are together equal (I. 13.) to two right angles; and CEA, AED have been demonstrated to be equal to two right angles; wherefore the angles CEA, AED are equal to the angles AED, DEB. Take away the common angle AED, and the remaining angle CEA is equal (Ax. 3.) to the remaining angle DEB. In the same manner it can be demonstrated that the angles CEB, AED are equal. Therefore if two straight lines, &c. Q. E. D.

COR. 1. From this it is manifest, that, if two straight lines cut one another, the angles they make at the point where they cut, are together equal to four right angles.

COR. 2. And consequently that all the angles made by any number of lines meeting in one point, are together equal to four right angles.

PROP. XVI. THEOR.

If one side of a triangle be produced, the exterior angle is greater than either of the interior opposite angles.

Let ABC be a triangle, and let its side BC be produced to D, the exterior angle ACD is greater than either of the interior opposite angles CBA, BAC.

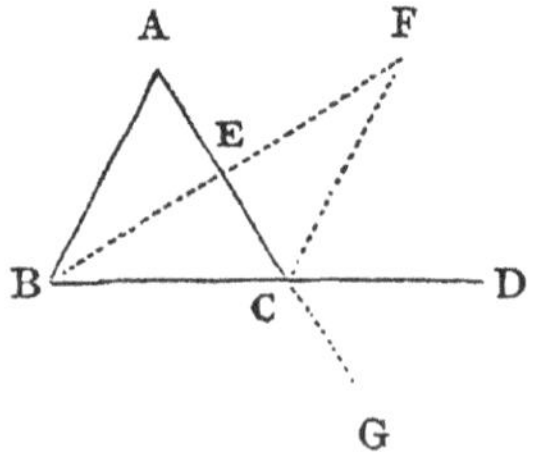

Bisect (I. 10.) AC in E, join BE and produce it to F, and make EF equal to BE; join also FC, and produce AC to G.

Because AE is equal to EC, and BE to EF; AE, EB are equal to CE, EF, each to each; and the angle AEB is equal (I. 15.) to the angle CEF, because they are opposite vertical angles; therefore the base AB is equal (I. 4.) to the base CF, and the triangle AEB to the triangle CEF, and the remaining angles to the remaining angles, each to each, to which the equal sides are opposite: wherefore the angle BAE is equal to the angle ECF; but the angle ECD is greater than the angle ECF; therefore the angle ACD is greater than BAE. In the same manner, if the side BC be bisected, it may be demonstrated that the angle BCG, that is (I. 15.), the angle ACD, is greater than the angle ABC. Therefore, if one side, &c. Q. E. D.

PROP. XVII. THEOR.

Any two angles of a triangle are together less than two right angles.

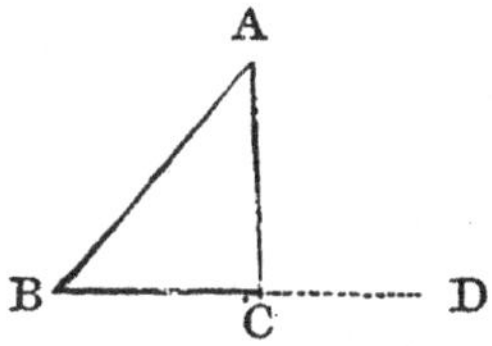

Let ABC be any triangle; any two of its angles together are less than two right angles.

Produce BC to D; and because ACD is the exterior angle of the triangle ABC, ACD is greater (I. 16.) than the interior and opposite angle ABC; to each of these add the angle ACB; therefore the angles ACD, ACB are greater than the angles ABC, ACB; but ACD, ACB are together equal (I. 13.) to two right angles; therefore the angles ABC, BCA are less than two right angles. In like manner, it may be demonstrated, that BAC, ACB, as also CAB, ABC, are less than two right angles. Therefore any two angles, &c. Q. E. D.

PROP. XVIII. THEOR.

The greater side of every triangle is opposite to the greater angle.

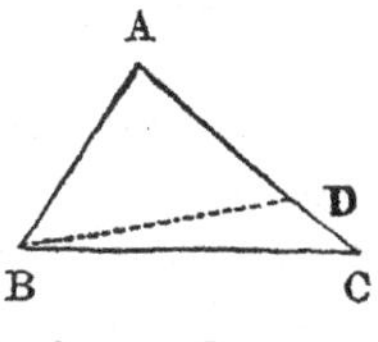

Let ABC be a triangle, of which the side AC is greater than the side AB; the angle ABC is also greater than the angle BCA.

Because AC is greater than AB, make (I. 3.) AD equal to AB, and join BD; and because ADB is the exterior angle of the triangle BDC, it is greater (I. 16.) than the interior and opposite angle DCB; but ADB is equal (I. 5.) to ABD, because the side AB is equal to the side AD; therefore the angle ABD is likewise greater than the angle ACB; wherefore much more is the angle ABC greater than ACB. Therefore the greater side, &c. Q.E.D.

Prop. XIX. Theor.

The greater angle of every triangle is subtended by the greater side, or has the greater side opposite to it.

Let ABC be a triangle, of which the angle ABC is greater than the angle BCA; the side AC is likewise greater than the side AB.

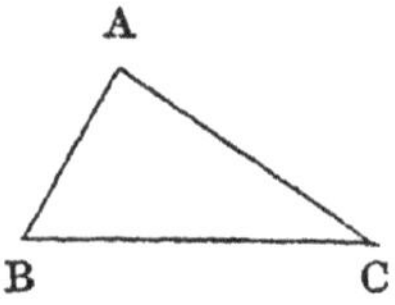

For, if it be not greater, AC must either be equal to AB, or less than it; it is not equal, because then the angle ABC would be equal (I. 5.) to the angle ACB; but it is not; therefore AC is not equal to AB; neither is it less; because then the angle ABC would be less (I. 18.) than the angle ACB; but it is not; therefore the side AC is not less than AB; and it has been shown that it is not equal to AB; therefore AC is greater than AB. Wherefore the angle, &c. Q.E.D.

Prop. XX. Theor.

Any two sides of a triangle are together greater than the third side.

Let ABC be a triangle; any two sides of it together are greater than the third side, viz. the sides BA, AC greater than the side BC; and AB, BC greater than AC; and BC, CA greater than AB.

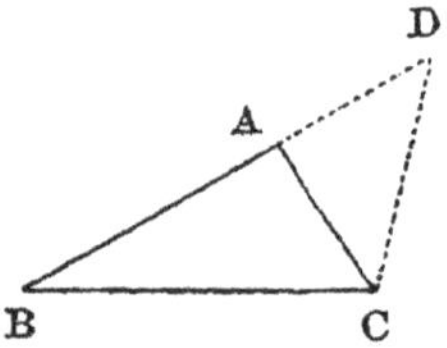

Produce BA to the point D, and make (I. 3.) AD equal to AC; and join DC.

Because DA is equal to AC, the angle ADC is likewise equal (I. 5.) to ACD; but the angle BCD is greater than the angle ACD; therefore the angle BCD is greater than the angle ADC; and because the angle BCD of the triangle DCB is greater than its angle BDC, and that the greater (I. 19.) side is opposite to the greater angle; therefore the side DB is greater than the side BC; but DB is equal to BA and AC; therefore the sides BA, AC are greater than BC. In the same manner it may be demonstrated, that the sides AB, BC are greater than CA, and BC, CA greater than AB. Therefore any two sides, &c. Q.E.D.

PROP. XXI. THEOR.

If, from the ends of the side of a triangle, there be drawn two straight lines to a point within the triangle, these shall be less than the other two sides of the triangle, but shall contain a greater angle.

Let the two straight lines BD, CD be drawn from B, C, the ends of the side BC of the triangle ABC, to the point D within it; BD and DC are less than the other two sides BA, AC of the triangle, but contain an angle BDC greater than the angle BAC.

Produce BD to E; and because two sides of a triangle are greater than the third side, the two sides BA, AE of the triangle ABE are greater than BE, to each of these add EC; therefore the sides BA, AC are greater than BE, EC: Again, because the two sides CE, ED of the triangle CED are greater than CD, add DB to each of these; therefore the sides CE, EB are greater than CD, DB; but it has been shown that BA, AC are greater than BE, EC; much more then are BA, AC greater than BD, DC.

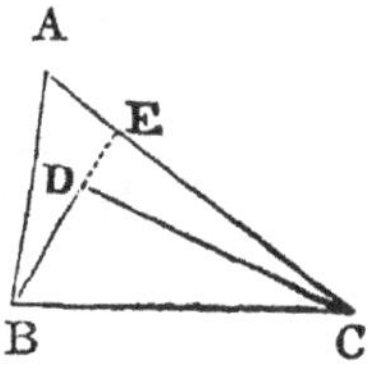

Again, because the exterior angle of a triangle is greater than the interior and opposite angle (I. 16.), the exterior angle BDC of the triangle CDE is greater than CED; for the same reason, the exterior angle CEB of the triangle ABE is greater than BAC; and it has been demonstrated that the angle BDC is greater than the angle CEB; much more then is the angle BDC greater than the angle BAC. Therefore, if from the ends of, &c. Q. E. D.

PROP. XXII. PROB.

To make a triangle of which the sides shall be equal to three given straight lines, but any two whatever of these must be greater than the third. (I. 20.)

Let A, B, C be the three given straight lines, of which any two whatever are greater than the third, viz. A and B greater than C; A and C greater than B; and B and C than A. It is required to make a triangle of which the sides shall be equal to A, B, C, each to each.

Take a straight line DE terminated at the point D, but unlimited towards E, and make (I. 3.) DF equal to A, FG to B, and GH equal to C; and from the centre F, at a distance FD, describe (Post. 3.) the circle DKL; and from the centre G, at the distance GH, describe (Post. 3.) another circle HLK; and join KF, KG; the triangle KFG has its sides equal to the three straight lines, A, B, C.

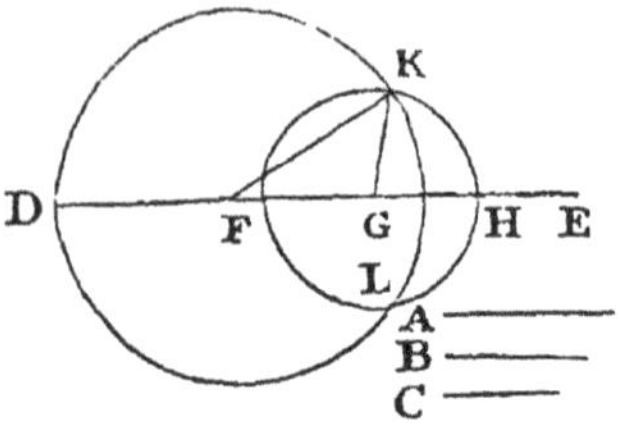

Because the point F is the centre of the circle DKL, FD is equal (Def. 15.) to FK; but FD is equal to the straight line A; therefore FK is equal to A: Again, because G is the centre of the circle LKH, GH is equal (Def. 15.) to GK; but GH is equal to C; therefore also GK is equal to C; and FG is equal to B; therefore the three straight lines KF, FG, GK, are equal to the three A, B, C: And therefore the triangle KFG has its three sides KF, FG, GK equal to the three given straight lines, A, B, C. Which was to be done.

Prop. XXIII. Prob

At a given point in a given straight line, to make a rectilineal angle equal to a given rectilineal angle.

Let AB be the given straight line, and A the given point in it, and DCE the given rectilineal angle; it is required to make an angle at the given point A in the given straight line AB, that shall be equal to the given rectilineal angle DCE.

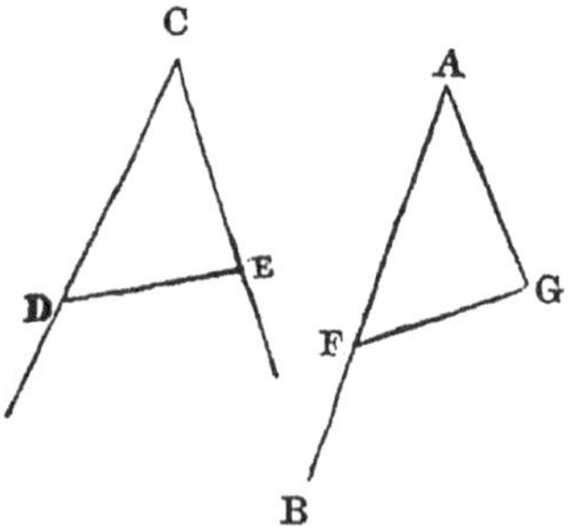

Take in CD, CE any points D, E, and join DE; and make (I. 22.) the triangle AFG the sides of which shall be equal to the three straight lines CD, DE, CE, so that CD be equal to AF, CE to AG, and DE to FG; and because DC, CE are equal to FA, AG, each to each, and the base DE to the base FG; the angle DCE is equal (I. 8.) to the angle FAG. Therefore, at the given point A in the given straight line AB, the angle FAG is made equal to the given rectilineal angle DCE. Which was to be done.

PROP. XXIV. THEOR.

If two triangles have two sides of the one equal to two sides of the other, each to each, but the angle contained by the two sides of one of them greater than the angle contained by the two sides equal to them, of the other; the base of that which has the greater angle shall be greater than the base of the other.

Let ABC, DEF be two triangles which have the two sides AB, AC equal to the two DE, DF, each to each, viz. AB equal to DE, and AC to DF; but the angle BAC greater than the angle EDF; the base BC is also greater than the base EF.

Of the two sides DE, DF, let DE be the side which is not greater than the other, and at the point D, in the straight line DE, make (I. 23.) the angle EDG equal to the angle BAC; and make DG equal (I. 3.) to AC or DF, and join EG, GF.

Because AB is equal to DE, and AC to DG, the two sides BA, AC are equal to the two ED, DG, each to each, and the angle BAC is equal to the angle EDG; therefore the base BC is equal (I. 4.) to the base EG; and because DG is equal to DF, the angle DFG is equal (I. 5.) to the angle DGF; but the angle DGF is greater than the angle EGF, therefore the angle DFG is greater than EGF; and much more is the angle EFG greater than the angle EGF; and because the angle EFG of the triangle EFG is greater than its angle EGF, and that the greater (I. 19.) side is opposite to the greater angle; the side EG is therefore greater than the side EF; but EG is equal to BC; and therefore also BC is greater than EF. Therefore, if two triangles, &c. Q.E.D.

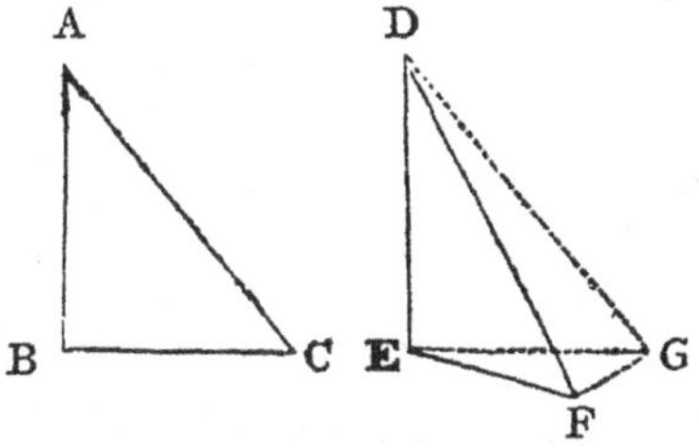

PROP. XXV. THEOR.

If two triangles have two sides of the one equal to two sides of the other, each to each, but the base of the one greater than the base of the other; the angle also contained by the sides of that which has the greater base, shall be greater than the angle contained by the sides equal to them, of the other.

Let ABC, DEF be two triangles which have the two sides

AB, AC equal to the two sides DE, DF, each to each, viz. AB equal to DE, and AC to DF; but the base CB is greater than the base EF; the angle BAC is likewise greater than the angle EDF.

For, if it be not greater, it must either be equal to it, or less; but the angle BAC is not equal to the angle EDF, because then the base BC would be equal (I. 4.) to EF; but it is not, therefore the angle BAC is not equal to the angle EDF; neither is it less; because then the base BC would be less (I. 24.) than the base EF; but it is not; therefore the angle BAC is not less than the angle EDF; and it was shown that it is not equal to it; therefore the angle BAC is greater than the angle EDF. Wherefore if two triangles, &c. Q.E.D.

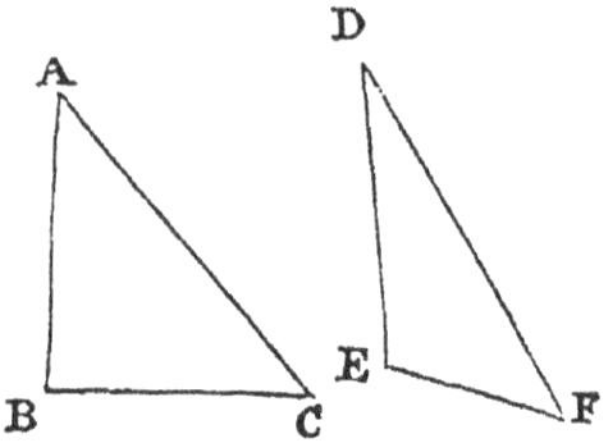

Prop. XXVI. Theor.

If two triangles have two angles of one equal to two angles of the other, each to each; and one side equal to one side, viz. either the sides adjacent to the equal angles, or the sides opposite to equal angles in each; then shall the other sides be equal each to each: and also the third angle of the one to the third angle of the other.

Let ABC, DEF be two triangles which have the angles ABC, BCA equal to the angles DEF, EFD, viz. ABC to DEF, and BCA to EFD; also one side equal to one side; and first let those sides be equal which are adjacent to the angles that are equal in the two triangles, viz. BC to EF; the other sides shall be equal, each to each, viz. AB to DE, and AC to DF; and the third angle BAC to the third angle EDF.

For, if AB be not equal to DE, one of them must be the greater. Let AB be the greater of the two, and make BG equal to DE, and join GC; therefore, because BG is equal to DE, and BC to EF (Hyp.), the two sides GB, BC are equal to the two DE, EF, each to each; and the angle GBC is equal to the angle DEF; therefore the base GC is equal (I. 4.) to the base DF, and the triangle GBC to the triangle DEF,

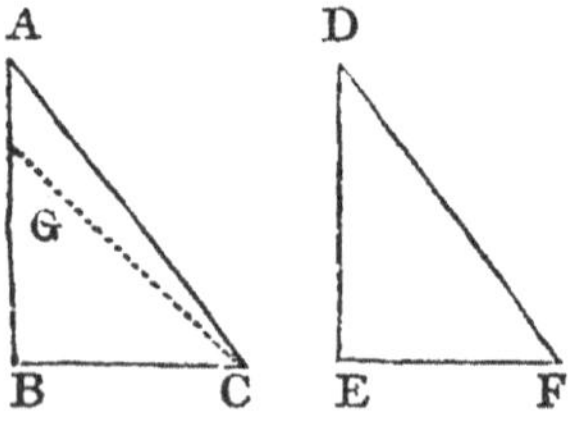

es AB, CD, make the alternate angles AEF, EFD equal to e another; AB is parallel to CD.

For, if it be not parallel, AB and CD being produced, shall et either towards B, D, or towards A, C; let them be proced and meet towards B, D in the point G; therefore GEF a triangle, and its exterior angle AEF is greater (I. 16.) n the interior and opposite gle EFG; but it is also equal (Hyp.), which is impossible: refore AB and CD being proed do not meet towards B, D. ike manner it may be demonated that they do not meet towards AC; but those straight s which meet neither way, though produced ever so far, parallel (Def. 35.) to one another. AB therefore is allel to CD. Wherefore if a straight line, &c. Q.E.D.

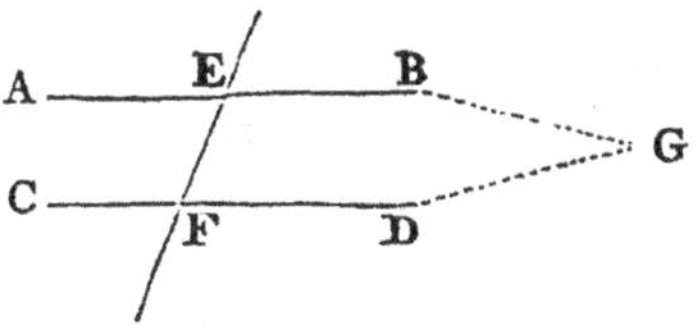

PROP. XXVIII. THEOR.

a straight line falling upon two other straight lines es the exterior angle equal to the interior and opposite the same side of the line; or makes the interior angles the same side together equal to two right angles; the straight lines shall be parallel to one another.

et the straight line EF, which falls the two straight lines AB, CD, make xterior angle EGB equal to the inr and opposite angle GHD upon the side; or make the interior angles he same side BGH, GHD together to two right angles; AB is parallel .

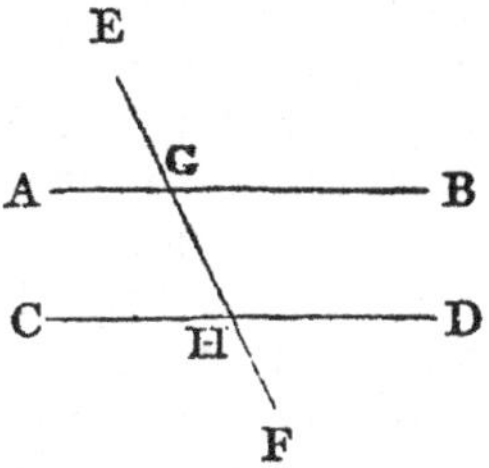

cause the angle EGB is equal to the GHD, and the angle EGB equal (I. 15.) to the angle the angle AGH is equal to the angle GHD, and they are lternate angles; therefore AB is parallel (I. 27.) to CD. n, because the angles BGH, GHD are equal (by Hyp.) to ight angles, and that AGH, BGH are also equal (I. 13.) right angles, the angles AGH, BGH are equal to the s BGH, GHD: Take away the common angle BGH; ore the remaining angle AGH is equal to the remaining GHD, and they are alternate angles; therefore AB is el to CD (I. 27.). Wherefore if a straight line, &c. Q.E.D.

and the other angles to the other angles, which the equal sides are opposite; therefor is equal to the angle DFE; but DFE is, by equal to the angle BCA; wherefore aso is equal to the angle BCA, the less to the impossible; therefore AB is not unequal to equal to it; and BC is equal to EF: there BC are equal to the two DE, EF, each to ea ABC is equal to the angle DEF; the bas equal (I. 4.) to the base DF, and the third third angle EDF.

Next, let the sides which are opposite to equal angles in each triangle be equal to one another, viz. AB to DE; likewise in this case, the other sides shall be equal, AC to DF, and BC to EF; and also the third angle BAC to the third EDF.

For, if BC be not equal to EF, let BC them, and make BH equal to EF, and joi BH is equal to EF, and AB to DE (Hyp.), equal to the two DE, EF, each to each equal angles (Hyp.); therefore the base base DF, and the triangle ABH to the tr other angles shall be equal, each to each, sides are opposite; therefore the angle angle EFD; but EFD is equal to the therefore also the angle BHA is equal to is, the exterior angle BHA of the triang interior and opposite angle BCA, which wherefore BC is not unequal to EF, th and AB is equal to DE (Hyp.); therefo equal to the two DE, EF, each to ea equal angles; wherefore the base AC DF, and the third angle BAC to the thi fore if two triangles, &c. Q. E. D.

PROP. XXVII. T

If a straight line falling upon tu makes the alternate angles equal to straight lines shall be parallel.

Let the straight line EF, which fall

and the other angles to the other angles, each to each, to which the equal sides are opposite; therefore the angle GCB is equal to the angle DFE; but DFE is, by the hypothesis, equal to the angle BCA; wherefore aso the angle BCG is equal to the angle BCA, the less to the greater, which is impossible; therefore AB is not unequal to DE, that is, it is equal to it; and BC is equal to EF: therefore the two AB, BC are equal to the two DE, EF, each to each; and the angle ABC is equal to the angle DEF; the base therefore AC is equal (I. 4.) to the base DF, and the third angle BAC to the third angle EDF.

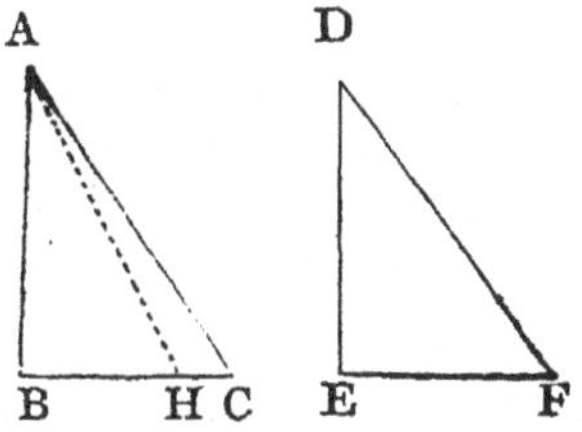

Next, let the sides which are opposite to equal angles in each triangle be equal to one another, viz. AB to DE; likewise in this case, the other sides shall be equal, AC to DF, and BC to EF; and also the third angle BAC to the third EDF.

For, if BC be not equal to EF, let BC be the greater of them, and make BH equal to EF, and join AH; and because BH is equal to EF, and AB to DE (Hyp.), the two AB, BH are equal to the two DE, EF, each to each; and they contain equal angles (Hyp.); therefore the base AH is equal to the base DF, and the triangle ABH to the triangle DEF, and the other angles shall be equal, each to each, to which the equal sides are opposite; therefore the angle BHA is equal to the angle EFD; but EFD is equal to the angle BCA (Hyp.); therefore also the angle BHA is equal to the angle BCA, that is, the exterior angle BHA of the triangle AHC is equal to its interior and opposite angle BCA, which is impossible (I. 16.); wherefore BC is not unequal to EF, that is, it is equal to it; and AB is equal to DE (Hyp.); therefore the two AB, BC are equal to the two DE, EF, each to each; and they contain equal angles; wherefore the base AC is equal to the base DF, and the third angle BAC to the third angle EDF. Therefore if two triangles, &c. Q.E.D.

Prop. XXVII. Theor.

If a straight line falling upon two other straight lines makes the alternate angles equal to one another, these two straight lines shall be parallel.

Let the straight line EF, which falls upon the two straight

lines AB, CD, make the alternate angles AEF, EFD equal to one another; AB is parallel to CD.

For, if it be not parallel, AB and CD being produced, shall meet either towards B, D, or towards A, C; let them be produced and meet towards B, D in the point G; therefore GEF is a triangle, and its exterior angle AEF is greater (I. 16.) than the interior and opposite angle EFG; but it is also equal to it (Hyp.), which is impossible: therefore AB and CD being produced do not meet towards B, D. In like manner it may be demonstrated that they do not meet towards AC; but those straight lines which meet neither way, though produced ever so far, are parallel (Def. 35.) to one another. AB therefore is parallel to CD. Wherefore if a straight line, &c. Q.E.D.

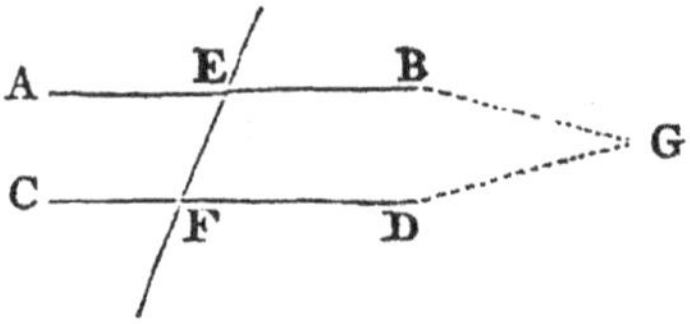

PROP. XXVIII. THEOR.

If a straight line falling upon two other straight lines makes the exterior angle equal to the interior and opposite upon the same side of the line; or makes the interior angles upon the same side together equal to two right angles; the two straight lines shall be parallel to one another.

Let the straight line EF, which falls upon the two straight lines AB, CD, make the exterior angle EGB equal to the interior and opposite angle GHD upon the same side; or make the interior angles on the same side BGH, GHD together equal to two right angles; AB is parallel to CD.

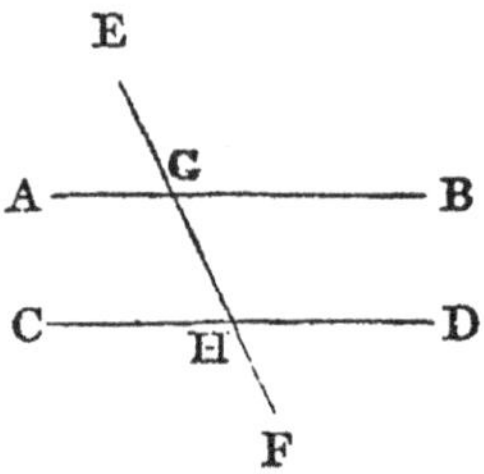

Because the angle EGB is equal to the angle GHD, and the angle EGB equal (I. 15.) to the angle AGH, the angle AGH is equal to the angle GHD, and they are the alternate angles; therefore AB is parallel (I. 27.) to CD. Again, because the angles BGH, GHD are equal (by Hyp.) to two right angles, and that AGH, BGH are also equal (I. 13.) to two right angles, the angles AGH, BGH are equal to the angles BGH, GHD: Take away the common angle BGH; therefore the remaining angle AGH is equal to the remaining angle GHD, and they are alternate angles; therefore AB is parallel to CD (I. 27.). Wherefore if a straight line, &c. Q.E.D.

PROP. XXIX. THEOR.

If a straight line fall upon two parallel straight lines, it makes the alternate angles equal to one another; and the exterior angle equal to the interior and opposite upon the same side; and likewise the two interior angles upon the same side together equal to two right angles.

Let the straight line EF fall upon the parallel straight lines AB, CD; the alternate angles AGH, GHD are equal to one another; and the exterior angle EGB is equal to the interior and opposite, upon the same side, GHD; and the two interior angles BGH, GHD upon the same side are together equal to two right angles.

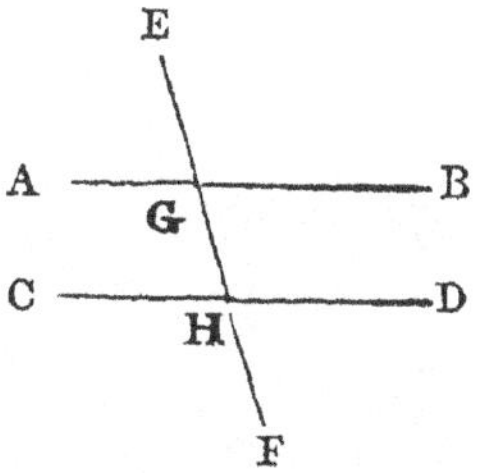

For, if AGH be not equal to GHD, one of them must be greater than the other; let AGH be the greater; and because the angle AGH is greater than the angle GHD, add to each of them the angle BGH; therefore the angles AGH, BGH are greater than the angles BGH, GHD; but the angles AGH, BGH are equal (I. 13.) to two right angles; therefore the angles BGH, GHD are less than two right angles; but those straight lines which, with another straight line falling upon them, make the interior angles on the same side less than two right angles, do meet (Ax. 12.) together if continually produced; therefore the straight lines AB, CD, if produced far enough, shall meet; but they never meet, since they are parallel by the hypothesis; therefore the angle AGH is not unequal to the angle GHD, that is, it is equal to it; but the angle AGH is equal (I. 15.) to the angle EGB; therefore likewise EGB is equal to GHD; add to each of these the angle BGH; therefore the angles EGB, BGH are equal to the angles BGH, GHD; but EGB, BGH are equal (I. 13.) to two right angles; therefore also BGH, GHD are equal to two right angles. Wherefore, if a straight line, &c. Q.E.D.

PROP. XXX. THEOR.

Straight lines which are parallel to the same straight line parallel to one another.

Let AB, CD be each of them parallel to EF; AB is also parallel to CD.

Let the straight line GHK cut AB, EF, CD; and because GHK cuts the parallel straight lines AB, EF, the angle AGH is equal (I. 29.) to the angle GHF. Again, because the straight line GHK cuts the parallel straight lines EF, CD, the angle GHF is equal (I. 29.) to the angle GKD; and it was shown that the angle AGK is equal to the angle GHF; therefore also AGK is equal to GKD; and they are alternate angles; therefore AB is parallel (I. 27.) to CD. Wherefore straight lines, &c. Q.E.D.

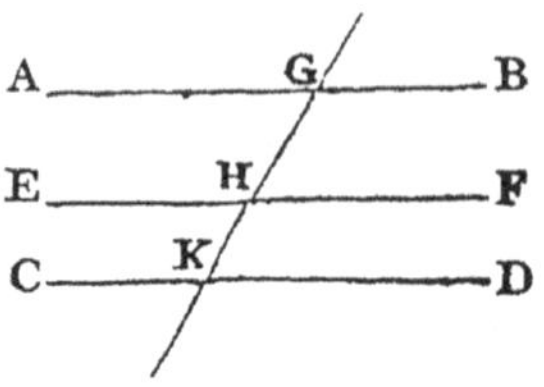

PROP. XXXI. PROB.

To draw a straight line through a given point parallel to a given straight line.

Let A be the given point, and BC the given straight line; it is required to draw a straight line through the point A, parallel to the straight line BC.

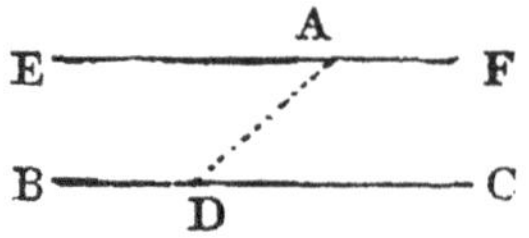

In BC take any point D, and join AD; and at the point A, in the straight line AD, make (I. 23.) the angle DAE equal to the angle ADC; and produce the straight line EA to F.

Because the straight line AD, which meets the two straight lines BC, EF, makes the alternate angles EAD, ADC equal to one another, EF is parallel (I. 27.) to BC. Therefore the straight line EAF is drawn through the given point A parallel to the given straight line BC. Which was to be done.

PROP. XXXII. THEOR.

If a side of any triangle be produced, the exterior angle is equal to the two interior and opposite angles; and the three interior angles of every triangle are equal to two right angles.

Let ABC be a triangle, and let one of its sides BC be produced to D; the exterior angle ACD is equal to the two interior and opposite angles CAB, ABC, and the three interior angles of the triangles, viz. ABC, BCA, CAB, are together equal to two right angles.

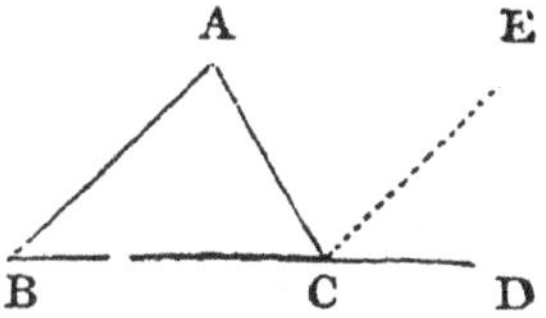

Through the point C draw CE parallel (I. 31.) to the straight line AB; and because AB is parallel to CE, and AC meets them, the alternate angles BAC, ACE are equal (I. 29.). Again, because AB is parallel to CE, and BD falls upon them, the exterior angle ECD is equal to the interior and opposite angle ABC; but the angle ACE was shown to be equal to the angle BAC; therefore the whole exterior angle ACD is equal to the two interior and opposite angles CAB, ABC; to these equals add the angle ACB, and the angles ACD, ACB are equal to the three angles CBA, BAC, ACB; but the angles ACD, ACB are equal (I. 13.) to two right angles; therefore also the angles CBA, BAC, ACB are equal to two right angles. Wherefore if a side of a triangle, &c. Q.E.D.

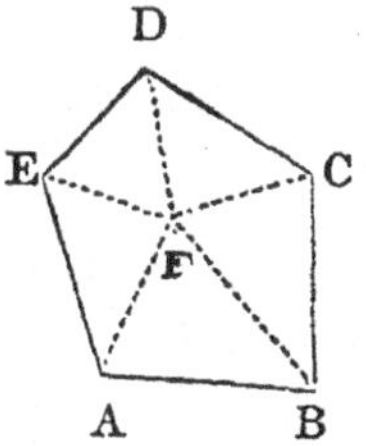

COR. 1. All the interior angles of any rectilineal figure, together with four right angles, are equal to twice as many right angles as the figure has sides.

For any rectilineal figure ABCDE can be divided into as many triangles as the figure has sides, by drawing straight lines from a point F within the figure to each of its angles. And, by the preceding proposition, all the angles of these triangles are equal to twice as many right angles as there are triangles, that is, as there are sides of the figure; and the same angles are equal to the angles of the figure, together with the angles at the point F, which is the common vertex of the triangles; that is (Cor. 2. I. 15.), together with four right angles. Therefore all the angles of the figure, together with four right angles, are equal to twice as many right angles as the figure has sides.

COR. 2. All the exterior angles of any rectilineal figure, are together equal to four right angles.

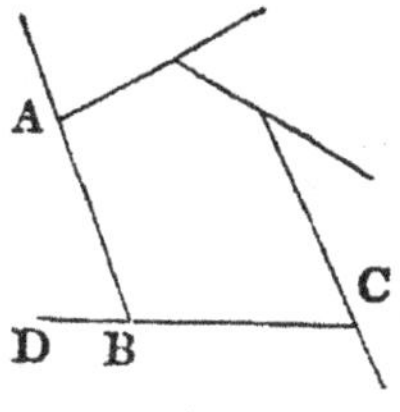

Because every interior angle ABC, with its adjacent exterior ABD, is equal (I. 13.) to two right angles; therefore all the interior, together with all the exterior angles of the figure, are equal to twice as many right angles as there are sides of the figure; that is, by the foregoing corollary, they are equal to all the interior angles of the figure, together with four right angles; therefore all the exterior angles are equal to four right angles.

Prop. XXXIII. Theor.

The straight lines which join the extremities of two equal and parallel straight lines, towards the same parts, are also themselves equal and parallel.

Let AB, CD be equal and parallel straight lines, and joined towards the same parts by the straight lines AC, BD; AC, BD are also equal and parallel.

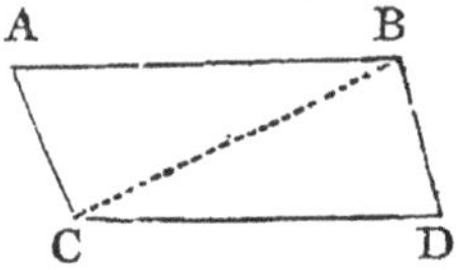

Join BC; and because AB is parallel to CD, and BC meets them, the alternate angles ABC, BCD are equal (I. 29.); and because AB is equal to CD (Hyp.), and BC common to the two triangles ABC, DCB, the two sides AB, BC are equal to the two DC, CB; and the angle ABC is equal to the angle BCD; therefore the base AC is equal (I. 4.) to the base BD, and the triangle ABC to the triangle BCD, and the other angles to the other angles (I. 4.), each to each, to which the equal sides are opposite; therefore the angle ACB is equal to the angle CBD; and because the straight line BC meets the two straight lines AC, BD, and makes the alternate angles ACB, CBD equal to one another, AC is parallel (I. 27.) to BD; and it was shown to be equal to it. Therefore, straight lines, &c. Q.E.D.

Prop. XXXIV. Theor.

The opposite sides and angles of parallelograms are equal to one another, and the diameter bisects them, that is, divides them in two equal parts.

N. B. — A parallelogram is a four-sided figure, of which the opposite sides are parallel; and the diameter is the straight line joining two of its opposite angles.

Let ACDB be a parallelogram, of which BC is a diameter; the opposite sides and angles of the figure are equal to one another; and the diameter BC bisects it.

Because AB is parallel to CD, and BC meets them, the alternate angles ABC, BCD are equal (I. 29.) to one another; and because AC is parallel to BD, and BC meets them, the alternate angles ACB, CBD are equal (I. 29.) to one another; wherefore the two triangles ABC, CBD have two angles ABC,

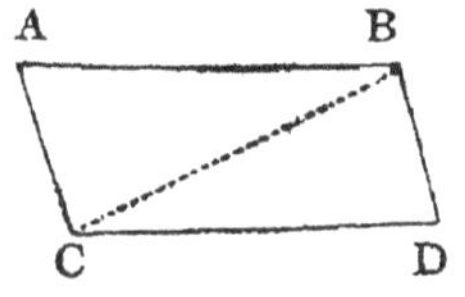

BCA in one, equal to two angles BCD, CBD in the other, each to each, and one side BC common to the two triangles, which is adjacent to their equal angles; therefore their other sides shall be equal, each to each, and the third angle of the one to the third angle of the other (I. 26.), viz. the side AB to the side CD, and AC to BD, and the angle BAC equal to the angle BDC: And because the angle ABC is equal to the angle BCD, and the angle CBD to the angle ACB, the whole angle ABD is equal to the whole angle ACD: And the angle BAC has been shown to be equal to the angle BDC; therefore the opposite sides and angles of parallelograms are equal to one another; also, their diameter bisects them; for AB being equal to CD, and BC common, the two AB, BC are equal to the two DC, CB, each to each; and the angle ABC is equal to the angle BCD; therefore the triangle ABC is equal (I. 4.) to the triangle BCD, and the diameter BC divides the parallelogram ACDB into two equal parts. Q. E. D.

PROP. XXXV. THEOR.

Parallelograms upon the same base and between the same parallels, are equal to one another.

Let the parallelograms ABCD, EBCF (see the 2d and 3d figures) be upon the same base BC, and between the same parallels AF, BC; the parallelogram ABCD shall be equal to the parallelogram EBCF.

If the sides AD, DF of the parallelograms ABCD, DBCF opposite to the base BC be terminated in the same point D; it is plain that each of the parallelograms is double (I. 34.) of the triangle BDC; and they are therefore equal to one another.

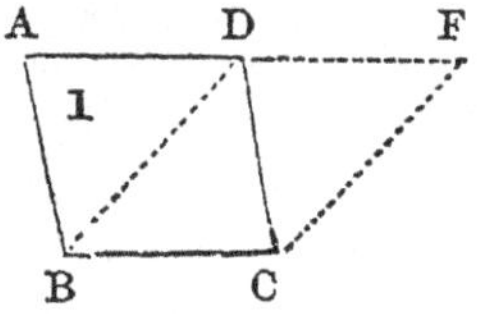

But, if the sides AD, EF, opposite to the base BC of the parallelograms ABCD, EBCF, be not terminated in the same point; then, because ABCD is a parallelogram, AD is equal (I. 34.) to

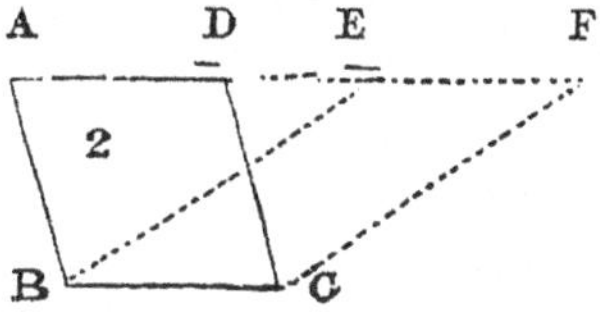

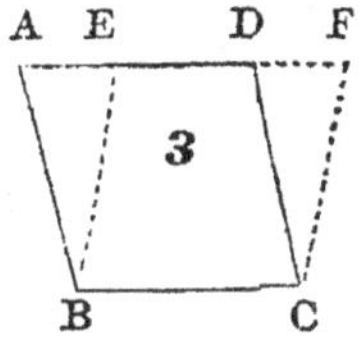

BC; for the same reason EF is equal to BC; wherefore AD is

equal (Ax. 1.) to EF; and DE is common; therefore the whole, or the remainder, AE is equal (Ax. 2. or 3.) to the whole, or the remainder DF; AB also is equal to DC; and the two EA, AB are therefore equal to the two FD, DC, each to each; and the exterior angle FDC is equal (I. 29.) to the interior EAB, therefore the base EB is equal to the base FC, and the triangle EAB equal (I. 4.) to the triangle FDC; take the triangle FDC from the trapezium ABCF, and from the same trapezium take the triangle EAB; the remainders, therefore, are equal (Ax. 3.), that is, the parallelogram ABCD is equal to the parallelogram EBCF. Therefore, parallelograms upon the same base, &c. Q. E. D.

PROP. XXXVI. THEOR.

Parallelograms upon equal bases, and between the same parallels, are equal to one another.

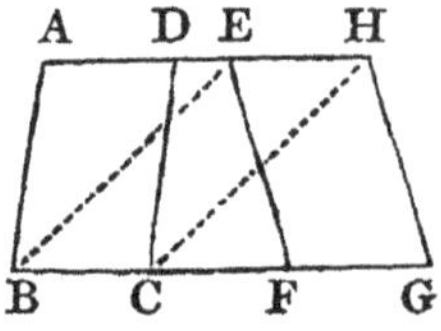

Let ABCD, EFGH be parallelograms upon equal bases BC, FG, and between the same parallels AH, BG; the parallelogram ABCD is equal to EFGH.

Join BE, CH; and because BC is equal to FG, and FG to (I. 34.) EH, BC is equal to EH; and they are parallels, and joined towards the same parts by the straight lines BE, CH: But straight lines which join equal and parallel straight lines towards the same parts, are themselves equal and parallel (I. 33.); therefore EB, CH are both equal and parallel, and EBCH is a parallelogram; and it is equal (I. 35.) to ABCD, because it is upon the same base BC, and between the same parallels BC, AD: For the like reason, the parallelogram EFGH is equal to the same EBCH: Therefore also the parallelogram ABCD is equal to EFGH. Wherefore parallelograms, &c.

PROP. XXXVII. THEOR.

Triangles upon the same base, and between the same parallels, are equal to one another.

Let the triangles ABC, DBC be upon the same base BC and between the same parallels AD, BC: The triangle ABC is equal to the triangle DBC.

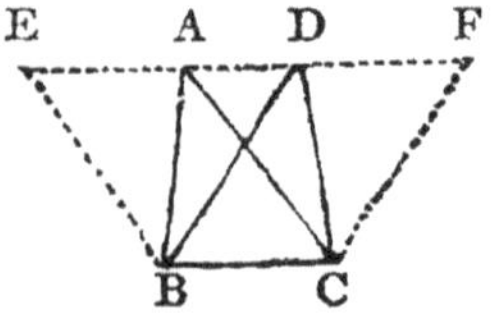

Produce AD both ways to the points E, F, and through B draw (I. 31.) BE parallel to CA; and through C draw CF parallel to BD: Therefore each of the figures EBCA, DBCF is a parallelogram; and EBCA is equal (I. 35.) to DBCF, because they are upon the same base BC, and between the same parallels BC, EF; and the triangle ABC is the half of the parallelogram EBCA, because the diameter AB bisects it (I. 34.); and the triangle DBC is the half of the parallelogram DBCF, because the diameter DC bisects it: But the halves of equal things are equal (Ax. 7.); therefore the triangle ABC is equal to the triangle DBC. Wherefore triangles, &c. Q. E. D.

Prop. XXXVIII. Theor.

Triangles upon equal bases, and between the same parallels, are equal to one another.

Let the triangles ABC, DEF be upon equal bases BC, EF, and between the same parallels BF, AD: The triangle ABC is equal to the angle DEF.

Produce AD both ways to the point G, H, and through B draw BG parallel (I. 31.) to CA, and through F draw FH parallel to ED: Then each of the figures GBCA, DEFH is a parallelogram; and they are equal to (I. 36.) one another, because they are upon equal bases BC, EF, and between the same parallels BF, GH; and the triangle ABC is the half (I. 34.) of the parallelogram GBCA, because the diameter AB bisects it; and the triangle DEF is the half (I. 34.) of the parallelogram DEH, because the diameter DF bisects it: But the halves of equal things are equal (Ax. 7.), therefore the triangle ABC is equal to the triangle DEF. Wherefore triangles, &c. Q.E.D.

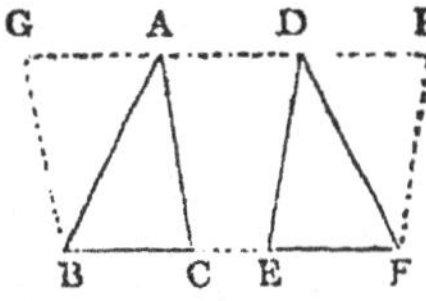

Prop. XXXIX. Theor.

Equal triangles upon the same base, and upon the same side of it, are between the same parallels.

Let the equal triangles ABC, DBC be upon the same base BC, and upon the same side of it; they are between the same parallels.

Join AD: AD is parallel to BC; for, if it is not, through

the point A draw (I. 31.) AE parallel to BC, and join EC: The triangle ABC is equal (I. 37.) to the triangle EBC, because it is upon the same base BC, and between the same parallels BC, AE: But the triangle ABC is equal to the triangle BDC (Hyp.); therefore also the triangle BDC is equal to the triangle EBC, the greater to the less, which is impossible: Therefore AE is not parallel to BC. In the same manner, it can be demonstrated that no other line but AD is parallel to BC; AD is therefore parallel to it. Wherefore equal triangles upon, &c. Q.E.D.

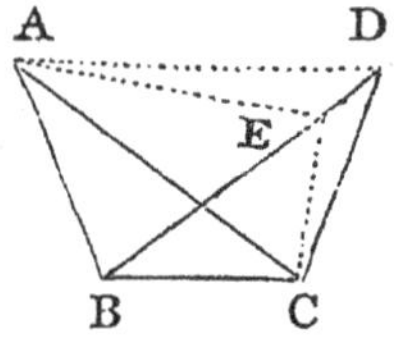

Prop. XL. Theor.

Equal triangles upon equal bases, in the same straight line, and towards the same parts, are between the same parallels.

Let the equal triangles ABC, DEF be upon equal bases BC, EF, in the same straight line BF, and towards the same parts; they are between the same parallels.

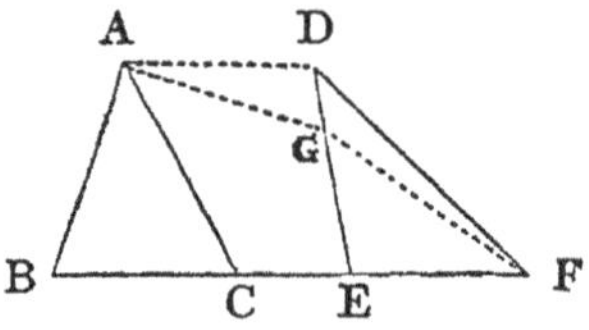

Join AD; AD is parallel to BC: For, if it is not, through A draw (I. 31.) AG parallel to BF, and join GF; the triangle ABC is equal (I. 38.) to the triangle GEF, because they are upon equal bases BC, EF, and between the same parallels BF, AG: But the triangle ABC is equal to the triangle DEF (Hyp.); therefore also the triangle DEF is equal to the triangle GEF, the greater to the less, which is impossible: Therefore AG is not parallel to BF: And in the same manner it can be demonstrated that there is no other parallel to it but AD; AD is therefore parallel to BF. Wherefore equal triangles, &c. Q. E. D.

Prop. XLI. Theor.

If a parallelogram and triangle be upon the same base, and between the same parallels; the parallelogram shall be double of the triangle.

Let the parallelogram ABCD and the triangle EBC be upon

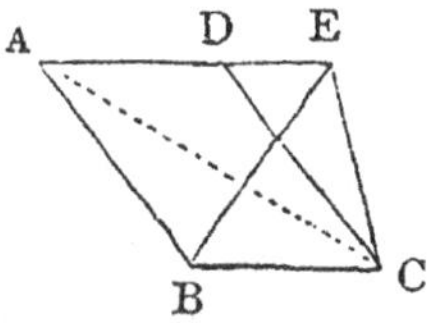

the same base BC, and between the same parallels BC, AE; the parallelogram ABCD is double of the triangle EBC.

Join AC; then the triangle ABC is equal (I. 37.) to the triangle EBC, because they are upon the same base BC, and between the same parallels BC, AE. But the parallelogram ABCD is double (I. 34.) of the triangle ABC, because the diameter AC divides it into two equal parts; wherefore ABCD is also double of the triangle EBC. Therefore if a parallelogram, &c. Q.E.D.

Prop. XLII. Prob.

To describe a parallelogram that shall be equal to a given triangle, and have one of its angles equal to a given rectilineal angle.

Let ABC be the given triangle, and D the given rectilineal angle. It is required to describe a parallelogram that shall be equal to the given triangle ABC, and have one of its angles equal to D.

Bisect (I. 10.) BC in E, join AE, and at the point E in the straight line EC make (I. 23.) the angle CEF equal to D; and through A draw (I. 31.) AG parallel to EC, and through C draw CG (I. 31.) parallel to EF: Therefore FECG is a parallelogram: And because BE is equal to EC, the triangle ABE is likewise equal (I. 38.) to the triangle AEC, since they are upon equal bases BE, EC, and between the same parallels BC, AG; therefore the triangle ABC is double of the triangle AEC. And the parallelogram FECG is likewise double (I. 41.) of the triangle AEC, because it is upon the same base, and between the same parallels: Therefore the same parallelogram FECG is equal to the triangle ABC, and it has one of its angles CEF equal to the given angle D: Wherefore there has been described a parallelogram FECG equal to a given triangle ABC, having one of its angles CEF equal to the given angle D. Which was to be done.

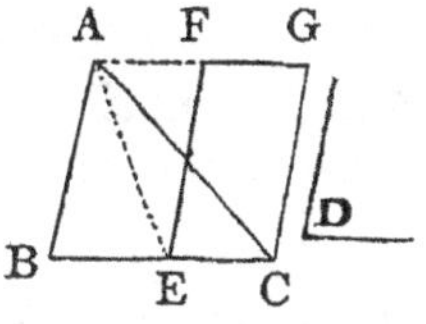

PROP. XLIII. THEOR.

The complements of the parallelogram which are about the diameter of any parallelogram, are equal to one another.

Let ABCD be a parallelogram, of which the diameter is AC, and EH, FG the parallelograms about AC, *that is, through which* AC *passes,* and BK, KD the other parallelograms which make up the whole figure ABCD, which are therefore called the complements: The complement BK is equal to the complement KD.

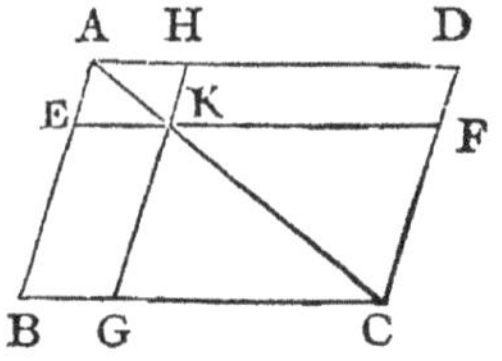

Because ABCD is a parallelogram, and AC its diameter, the triangle ABC is equal (I. 34.) to the triangle ADC: And, because EKHA is a parallelogram, the diameter of which is AK, the triangle AEK is equal to the triangle AHK: By the same reason, the triangle KGC is equal to the triangle KFC: Then, because the triangle AEK is equal to the triangle AHK, and the triangle KGC to KFC; the triangle AEK together with the triangle KGC is equal to the triangle AHK together with the triangle KFC: But the whole triangle ABC is equal to the whole ADC; therefore the remaining complement BK is equal to the remaining complement KD. Wherefore the complements, &c. Q.E.D.

PROP. XLIV. PROB.

To a given straight line to apply a parallelogram, which shall be equal to a given triangle, and have one of its angles equal to a given rectilineal angle.

Let AB be the given straight line, and C the given triangle, and D the given rectilineal angle. It is required to apply to the straight line AB a parallelogram equal to the triangle C, and having an angle equal to D.

Make (I. 42.) the parallelogram BEFG equal to the triangle C, and having the angle EBG equal to the angle D, so that BE be in the same straight line with AB, and produce FG to H; and through A draw (I. 31.) AH parallel to BG or EF, and join HB. Then because the

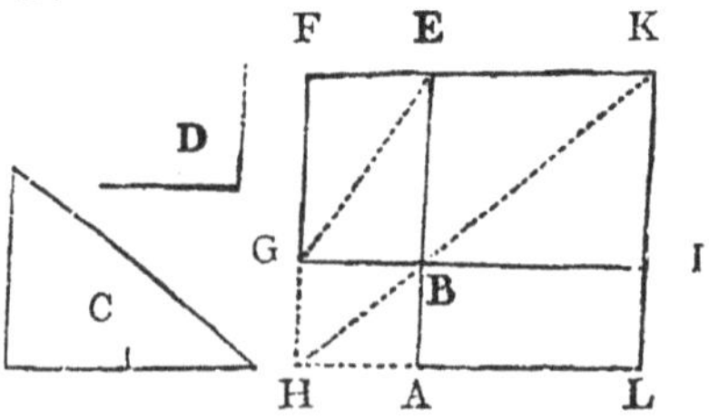

straight line HF falls upon the parallels AH, EF, the angles AHF, HFE, are together equal (I. 29.) to two right angles; wherefore the angles BHF, HFE are less than two right angles: But straight lines which with another straight line make the interior angles upon the same side less than two right angles, do meet (Ax. 12.) if produced far enough: Therefore HB, FE shall meet, if produced; let them meet in K, and through K draw KL parallel to EA or FH, and produce HA, GB to the points L, M: Then HLKF is a parallelogram, of which the diameter is HK, and AG, ME are the parallelograms about HK; and LB, BF are the complements; therefore LB is equal (I. 43.) to BF: But BF is equal to the triangle C; wherefore LB is equal to the triangle C; and because the angle GBE is equal (I. 15.) to the angle ABM, and likewise to the angle D (Constr.); the angle ABM is equal to the angle D: Therefore the parallelogram LB is applied to the straight line AB, is equal to the triangle C, and has the angle ABM equal to the angle D: Which was to be done.

PROP. XLV. PROB.

To describe a parallelogram equal to a given rectilineal figure, and having an angle equal to a given rectilineal angle.

Let ABCD be the given rectilineal figure, and E the given rectilineal angle. It is required to describe a parallelogram equal to ABCD, and having an angle equal to E.

Join DB, and describe (I. 42.) the parallelogram FH equal to the triangle ADB, and having the angle HKF equal to the angle E; and to the straight line GH apply (I. 44.) the parallelogram GM equal to the triangle DBC, having the angle GHM equal to the angle E: and because the angle E is equal to each of the angles FKH, GHM, the angle FKH is equal to GHM; add to each of these the angle KHG; therefore the angles FKH, KHG are equal to the angles KHG, GHM; but FKH, KHG are equal (I. 29.) to two right angles: therefore also KHG, GHM are equal to two right angles; and because at the point H in the straight line GH, the two straight lines KH, HM upon the opposite sides of it make the adjacent angles

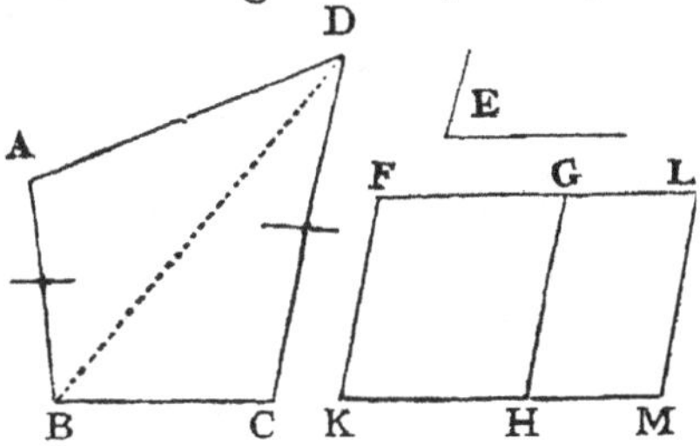

equal to two right angles, KH is in the same straight line (I. 14.) with HM; and because the straight line HG meets the parallels KM, FG, the alternate angles MHG, HGF are equal (I. 29.): Add to each of these the angle HGL: Therefore the angles MHG, HGL are equal to the angles HGF, HGL: But the angles MHG, HGL are equal (I. 29.) to two right angles; wherefore also the angles HGF, HGL are equal to two right angles, and FG is therefore in the same straight line with GL (I. 14.): And because KF is parallel to HG (Constr.), and HG to ML; KF is parallel (I. 30.) to ML: and KM, FL are parallels; wherefore KFLM is a parallelogram; and because the triangle ABD is equal to the parallelogram HF, and the triangle DBC to the parallelogram GM; the whole rectilineal figure ABCD is equal to the whole parallelogram KFLM; therefore the parallelogram KFLM has been described equal to the given rectilineal figure ABCD, having the angle FKM equal to the given angle E. Which was to be done.

COR. From this it is manifest how to a given straight line to apply a parallelogram, which shall have an angle equal to a given rectilineal angle, and shall be equal to a given rectilineal figure, viz. by applying (I. 44.) to the given straight line a parallelogram equal to the first triangle ABD, and having an angle equal to the given angle.

PROP. XLVI. PROB.

To describe a square upon a given straight line.

Let AB be the given straight line; it is required to describe a square upon AB.

From the point A draw (I. 11.) AC at right angles to AB; and make (I. 3.) AD equal to AB, and through the point D draw DE parallel (I. 31.) to AB, and through B draw BE parallel to AD; therefore ADEB is a parallelogram: whence AB is equal (I. 34.) to DE, and AD to BE: But BA is equal to AD; therefore the four straight lines BA, AD, DE, EB are equal to one another, and the parallelogram ADEB is equilateral, likewise all its angles are right angles; because the straight line AD meeting the parallels AB, DE, the angles BAD, ADE are equal (I. 29.) to two right angles; but BAD is a right angle; therefore also ADE is a right angle; but the opposite angles of paral

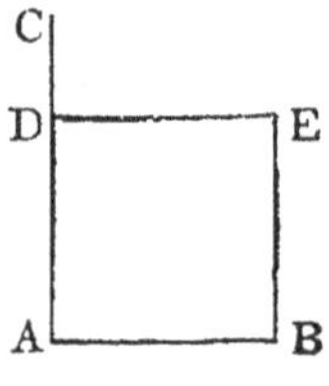

lelograms are equal (I. 34.); therefore each of the opposite angles ABE, BED is a right angle; wherefore the figure ADEB is rectangular, and it has been demonstrated that it is equilateral; it is therefore a square (Def. 30.), and it is described upon the given straight line AB. Which was to be done.

COR. Hence every parallelogram that has one right angle has all its angles right angles.

PROP. XLVII. THEOR.

In any right angled triangle, the square which is described upon the side subtending the right angle, is equal to the squares described upon the sides which contain the right angle.

Let ABC be a right angled triangle having the right angle BAC; the square described upon the side BC is equal to the squares described upon BA, AC.

On BC describe (I. 46.) the square BDEC, and on BA, AC the squares GB, HC; and through A draw (I. 31.) AL parallel to BD or CE, and join AD, FC; then, because each of the angles BAC, BAG is a right angle (Def. 30.), the two straight lines AC, AG upon the opposite sides of AB, make with it at the point A the adjacent angles equal to two right angles; therefore CA is in the same straight line (I. 14.) with AG; for the same reason, AB and AH are in the same straight line; and because the angle DBC is equal to the angle FBA, each of them being a right angle, add to each the angle ABC, and the whole

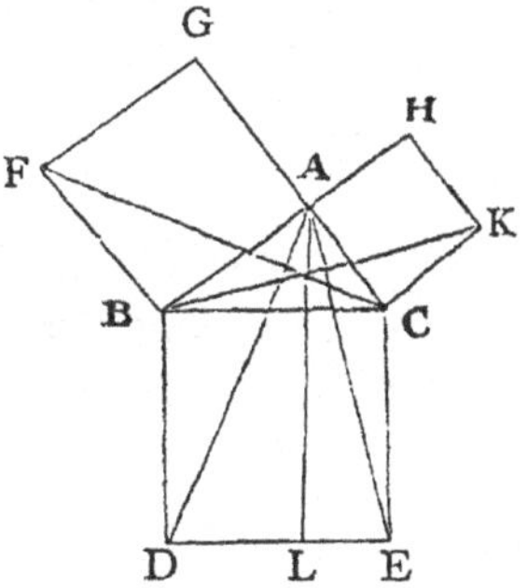

angle DBA is equal (Ax. 2.) to the whole FBC; and because the two sides AB, BD are equal to the two FB, BC, each to each, and the angle DBA equal to the angle FBC; therefore the base AD is equal (I. 4.) to the base FC, and the triangle ABD to the triangle FBC: Now the parallelogram BL is double (I. 41.) of the triangle ABD, because they are upon the same base BD, and between the same parallels, BD, AL; and the square GB is double of the triangle FBC, because these also are upon the same base FB, and between the same parallels FB, GC. But the doubles of equals are equal (Ax. 6.)

to one another: Therefore the parallelogram BL is equal to the square GB: And, in the same manner, by joining AE, BK, it is demonstrated that the parallelogram CL is equal to the square HC: Therefore the whole square BDEC is equal to the two squares GB, HC; and the square BDEC is described upon the straight line BC, and the squares GB, HC upon BA, AC: Wherefore the square upon the side BC is equal to the squares upon the sides BA, AC. Therefore, in any right angled triangle, &c. Q.E.D.

Prop. XLVIII. Theor.

If the square described upon one of the sides of a triangle, be equal to the squares described upon the other two sides of it; the angle contained by these two sides is a right angle.

If the square described upon BC, one of the sides of the triangle ABC, be equal to the squares upon the other sides BA, AC, the angle BAC is a right angle.

From the point A draw (I. 11.) AD at right angles to AC, and make AD equal to BA, and join DC: Then, because DA is equal to AB, the square of DA is equal to the square of AB: To each of these add the square of AC; therefore the squares of DA, AC, are equal to the squares of BA, AC: But the square of DC is equal (I. 47.) to the squares of DA, AC, because DAC is a right angle; and the square of BC, by hypothesis, is equal to the squares of BA, AC; therefore the square of DC is equal to the square of BC; and therefore also the side DC is equal to the side BC. And because the side DA is equal to AB, and AC common to the two triangles DAC, BAC, the two DA, AC are equal to the two BA, AC; and the base DC is equal to the base BC; therefore the angle DAC is equal (I. 8.) to the angle BAC: But DAC is a right angle; therefore also BAC is a right angle Therefore, if the square, &c. Q. E. D.

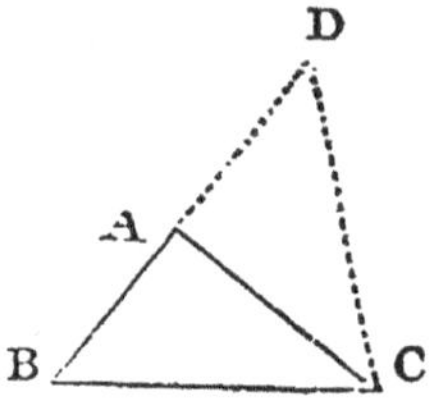

EUCLID'S

ELEMENTS OF GEOMETRY.

BOOK II.

DEFINITIONS.

I.

Every right angled parallelogram is said to be contained by any two of the straight lines which contain one of the right angles.

II.

In every parallelogram, any of the parallelograms about a diameter, together with the two complements, is called a Gnomon. "Thus the parallelogram HG, together with the complements AF, FC, is the gnomon, which is more briefly expressed by the letters AGK, or EHC, which are at the opposite angles of the parallelograms which make the gnomon."

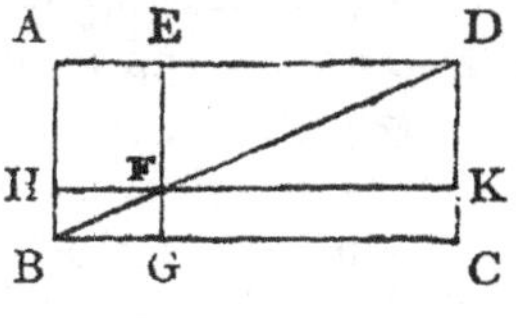

Prop. I. Theor.

If there be two straight lines, one of which is divided into any number of parts; the rectangle contained by the two straight lines, is equal to the rectangles contained by the undivided line, and the several parts of the divided line.

Let A and BC be two straight lines; and let BC be divided

into any parts in the points D, E; the rectangle contained by the straight lines A, BC is equal to the rectangle contained by A, BD, together with that contained by A, DE, and that contained by A, EC.

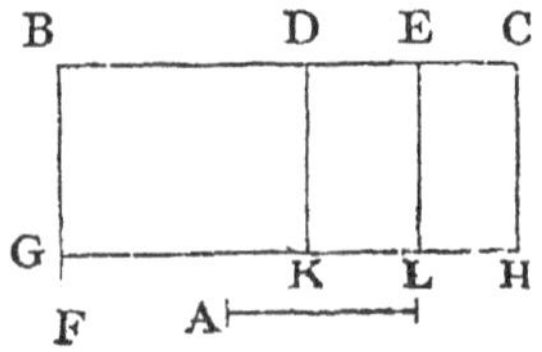

From the point B draw (I. 11.) BF at right angles to BC, and make BG equal (I. 3.) to A; and through G draw (I. 31.) GH parallel to BC; and through D, E, C, draw (I. 31.) DK, EL, CH parallel to BG; then the rectangle BH is equal to the rectangles BK, DL, EH; and BH is contained by A, BC, for it is contained by GB, BC, and GB is equal to A; and BK is contained by A, BD, for it is contained by GB, BD, of which GB is equal to A; and DL is contained by A, DE, because DK, that is (I. 34.) BG, is equal to A; and in like manner the rectangle EH is contained by A, EC: Therefore the rectangle contained by A, BC is equal to the several rectangles contained by A, BD, and by A, DE; and also by A, EC. Wherefore, if there be two straight lines, &c. Q.E.D.

PROP. II. THEOR.

If a straight line be divided into any two parts, the rectangles contained by the whole and each of the parts are together equal to the square of the whole line.

Let the straight line AB be divided into any two parts in the point C; the rectangle contained by AB, BC, together with the rectangle* AB, AC, shall be equal to the square of AB.

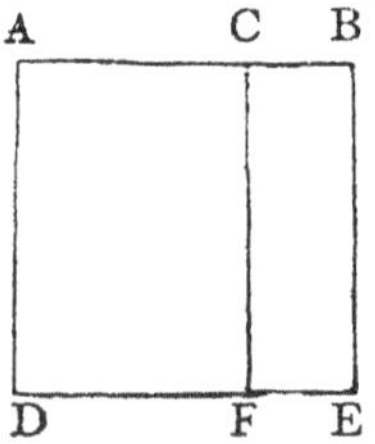

Upon AB describe (I. 46.) the square ADEB, and through C draw (I. 31.) CF parallel to AD or BE; then AE is equal to the rectangles AF, CE; and AE is the square of AB; and AF is the rectangle contained by BA, AC; for it is contained by DA, AC, of which AD is equal to AB; and CE is contained by AB, BC, for BE is equal to AB; therefore the rectangle contained by AB, AC together with the rectangle AB, BC, is equal to the square of AB. If therefore a straight line, &c. Q.E.D.

* N. B. — To avoid repeating the word *contained* too frequently, the rectangle contained by two straight lines AB, AC is sometimes simply called the rectangle AB, AC.

PROP. III. THEOR.

If a straight line be divided into any two parts, the rectangle contained by the whole and one of the parts is equal with the rectangle contained by the two parts, together with the square of the aforesaid part.

Let the straight line AB be divided into two parts in the point C; the rectangle AB, BC is equal to the rectangle AC, CB, together with the square of BC.

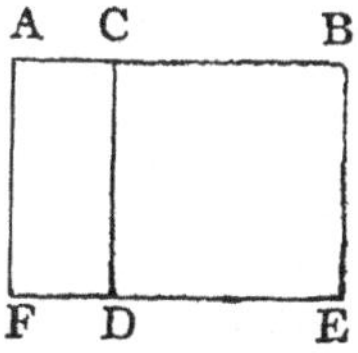

Upon BC describe (I. 46.) the square CDEB, and produce ED to F, and through A draw (I. 31.) AF parallel to CD or BE; then the rectangle AE is equal to the rectangles AD, CE; and AE is the rectangle contained by AB, BC, for it is contained by AB, BE, of which BE is equal to BC; and AD is contained by AC, CB, for CD is equal to CB; and DB is the square of BC; therefore the rectangle AB, BC is equal to the rectangle AC, CB together with the square of BC. If therefore a straight, &c. Q. E. D.

PROP. IV. THEOR.

If a straight line be divided into any two parts, the square of the whole line is equal to the squares of the two parts, together with twice the rectangle contained by the parts.

Let the straight line AB be divided into any two parts in C; the square of AB is equal to the squares of AC, CB, together with twice the rectangle contained by AC, CB.

Upon AB describe (I. 46.) the square ADEB, and join BD, and through C draw (I. 31.) CGF parallel to AD or BE, and through G draw HK parallel to AB or DE: And because CF is parallel to AD, and BD falls upon them, the exterior angle BGC is equal (I. 29.) to the interior and opposite angle ADB; but ADB is equal (I. 5.) to the angle ABD, because BA is equal to AD, being sides of a square; wherefore the angle

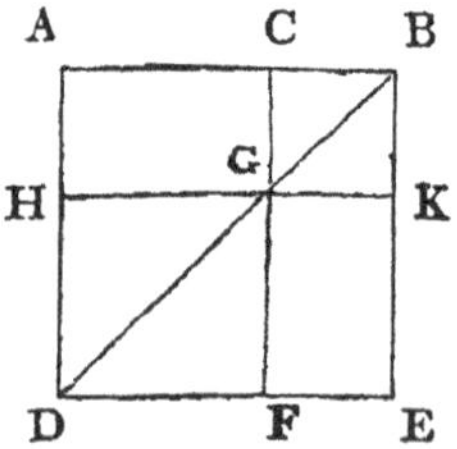

CGB is equal to the angle GBC; and therefore the side BC is equal (I. 6.) to the side CG: But CB is equal (I. 34.) also to GK, and CG to BK; wherefore the figure CGKB is equilateral: It is likewise rectangular; for CG is parallel to BK, and CB meets them; the angles KBC, GCB are therefore equal to two right angles; and KBC is a right angle; wherefore GCB is a right angle; and therefore also the angles (I. 34.) CGK, GKB opposite to these, are right angles, and CGKB is rectangular: But it is also equilateral, as was demonstrated; wherefore it is a square, and it is upon the side CB: For the same reason HF also is a square, and it is upon the side HG, which is equal to AC: therefore HF, CK are the squares of AC, CB; and because the complement AG is equal (I. 43.) to the complement GE, and that AG is the rectangle contained by AC, CB, for GC is equal to CB; Therefore GE is also equal to the rectangle AC, CB; wherefore AG, GE are equal to twice the rectangle AC, CB: And HF, CK are the squares of AC, CB; wherefore the four figures HF, CK, AG, GE are equal to the squares of AC, CB, and to twice the rectangle AC, CB: But HF, CK, AG, GE make up the whole figure ADEB, which is the square of AB: Therefore the square of AB is equal to the squares of AC, CB and twice the rectangle AC, CB. Wherefore if a straight line, &c. Q.E.D.

COR. From the demonstration, it is manifest that the parallelograms about the diameter of a square are likewise squares.

PROP. V. THEOR.

If a straight line be divided into two equal parts, and also into two unequal parts; the rectangle contained by the unequal parts, together with the square of the line between the points of section, is equal to the square of half the line.

Let the straight line AB be divided into two equal parts in the point C, and into two unequal parts at the point D; the rectangle AD, DB, together with the square of CD, is equal to the square of CB.

Upon CB describe (I. 46.) the square CEFB, join BE, and through D draw (I. 31.) DHG parallel to CE or BF; and through H draw KLM parallel to CB or EF; and also through

A draw AK parallel to CL or BM; and because the complement CH is equal (I. 43.) to the complement HF, to each of these add DM; therefore the whole CM is equal to the whole DF; but CM is equal (I. 36.) to AL, because AC is equal to CB; therefore also AL is equal to DF. To each of these add CH, and the whole AH is equal to DF and CH: But AH is the rectangle contained by AD, BD, for DH is equal (II. 4. Cor.) to DB; and DF, together with CH, is the gnomon CMG; therefore the gnomon CMG is equal to the rectangle AD, DB: To each of these add LG, which is equal (II. 4. Cor.) to the square of CD, therefore the gnomon GMG, together with LG, is equal to the rectangle AD, DB, together with the square of CD: But the gnomon CMG and LG make up the whole figure CEFB, which is the square of CB: Therefore the rectangle AD, DB, together with the square of CD, is equal to the square of CB. Wherefore, if a straight line, &c. Q. E. D.

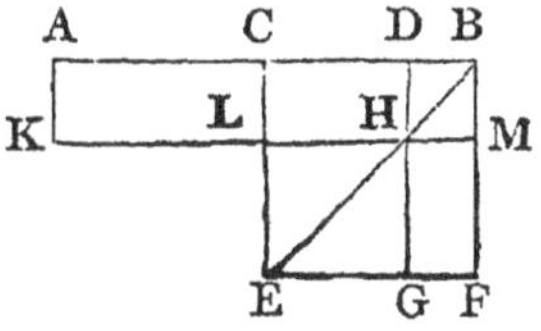

COR. From this proposition it is manifest, that the difference of the squares of two unequal lines AC, CD, is equal to the rectangle contained by their sum and difference.

PROP. VI. THEOR.

If a straight line be bisected, and produced to any point; the rectangle contained by the whole line thus produced, and the part of it produced, together with the square of half the line bisected, is equal to the square of the straight line, which is made up of the half and the part produced.

Let the straight line AB be bisected in C, and produced to the point D; the rectangle AD, DB, together with the square of CB, is equal to the square of CD.

Upon CD describe (I. 46.) the square CEFD, join DE, and through B draw (I. 31.) BHG parallel to CE or DF, and through H draw KLM parallel to AD or EF, and also through A draw AK parallel to CL or DM: and because AC is equal to CB, the rectangle AL is equal (I. 36.) to CH; but CH is equal (I. 43.) to HF; therefore also AL is equal to HF: To each of these add CM; therefore the whole AM is equal to the gnomon CMG: And AM is the rectangle con-

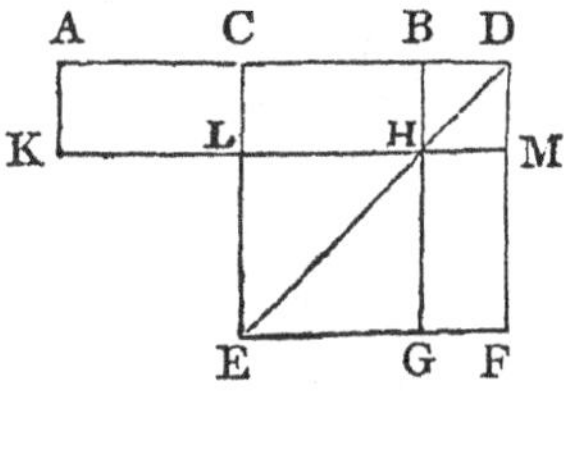

tained by AD, DB, for DM is equal (II. 4. Cor.) to DB: Therefore the gnomon CMG is equal to the rectangle AD, DB: Add to each of these LG, which is equal to the square of CB, therefore the rectangle AD, DB, together with the square of CB, is equal to the gnomon CMG and the figure LG: But the gnomon CMG and LG make up the whole figure CEFD, which is the square of CD; therefore the rectangle AD, DB, together with the square of CB, is equal to the square of CD. Wherefore, if a straight line, &c. Q.E.D.

Prop. VII. Theor.

If a straight line be divided into any two parts, the squares of the whole line, and one of the parts, are equal to twice the rectangle contained by the whole and that part, together with the square of the other part.

Let the straight line AB be divided into any two parts in the point C; the squares of AB, BC are equal to twice the rectangle AB, BC, together with the square of AC.

Upon AB describe (I. 46.) the square ADEB, and construct the figure as in the preceding propositions: And because AG is equal (I. 43.) to GE, add to each of them CK; the whole AK is therefore equal to the whole CE; therefore AK, CE are double of AK; but AK, CE are the gnomon AKF, together with the square CK, therefore the gnomon AKF, together with the square CK, is double of AK: But twice the rectangle AB, BC is double of AK, for BK is equal (II. 4. Cor.) to BC: Therefore the gnomon AKF, together with the square CK, is equal to twice the rectangle AB, BC: To each of these equals add HF, which is equal to the square of AC; therefore the gnomon AKF, together with the squares CK, HF, is equal to twice the rectangle AB, BC, and the square of AC: But the gnomon AKF, together with the squares CK, HF, make up the whole figure ADEB and CK, which are the squares of AB and BC: Therefore the squares of AB and BC are equal to twice the rectangle AB, BC, together with the squares of AC. Wherefore, if a straight line, &c. Q. E. D.

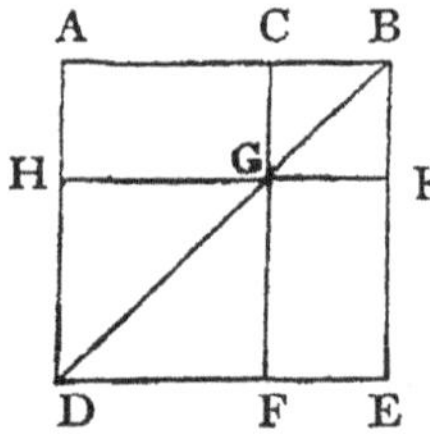

Prop. VIII. Theor.

If a straight line be divided into any two parts, four times the rectangle contained by the whole line, and one of the parts, together with the square of the other part, is equal to the square of the straight line which is made up of the whole and that part.

Let the straight line AB be divided into any two parts in the point C; four times the rectangle AB, BC, together with the square of AC, is equal to the square of the straight line made up of AB and BC together.

Produce AB to D, so that BD be equal to CB, and upon AD describe the square AEFD; and construct two figures such as in the preceding. Because CB is equal to BD, and that CB is equal (I. 34.) to GK, and BD to KN; therefore GK is equal to KN: For the same reason, PR is equal to RO; and because CB is equal to BD, and GK to KN, the rectangle CK is equal (I. 36.) to BN, and GR to RN: But CK is equal (I. 43.) to RN, because they are the complements of the parallelogram CO; therefore also BN is equal to GR, and the four rectangles BN, CK, GR, RN are therefore equal to one another, and so are quadruple of one of them CK: Again, because CB is equal to BD, and that BD is equal (II. 4. Cor.) to BK, that is, to CG; and CB equal to GK, that (II. 4. Cor.) is, to GP; therefore CG is equal to GP: and because CG is equal to GP, and PR to RO, the rectangle AG is equal to MP, and PL to RF: But MP is equal (I. 43.) to PL, because they are the complements of the parallelogram ML; wherefore AG is equal also to RF; Therefore the four rectangles AG, MP, PL, RF are equal to one another, and so are quadruple of one of them AG; and it was demonstrated, that the four CK, BN, GR, and RN are quadruple of CK: Therefore the eight rectangles which contain the gnomon AOH, are quadruple of AK: And because AK is the rectangle contained by AB, BC, for BK is equal to BC, therefore four times the rectangle AB, BC is quadruple of AK: But the gnomon AOH was demonstrated to be quadruple of AK; therefore four times the rectangle AB, BC is equal to the gnomon AOH. To each of these add XH, which

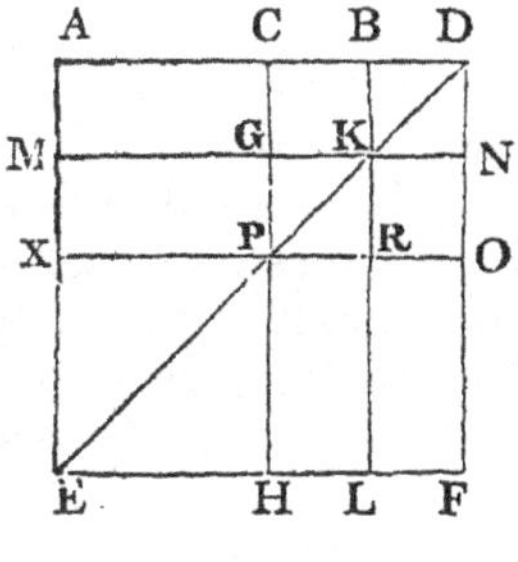

is equal (II. 4. Cor.) to the square of AC; Therefore four times the rectangle AB, BC, together with the square of AC, is equal to the gnomon AOH and the square XH: But the gnomon AOH and XH make up the figure AEFD, which is the square of AD: Therefore four times the rectangle AB, BC, together with the square of AC, is equal to the square of AD, that is, of AB and BC added together in one straight line. Wherefore, if a straight line, &c. Q. E. D.

PROP. IX. THEOR.

If a straight line be divided into two equal, and also into two unequal, parts; the squares of the two unequal parts are together double of the square of half the line, and of the square of the line between the points of section.

Let the straight line AB be divided at the point C into two equal, and at D into two unequal parts: The squares of AD, DB are together double of the squares of AC, CD.

From the point C draw (I. 11.) CE at right angles to AB, and make it equal to AC or CB, and join EA, EB; through D draw (I. 31.) DF parallel to CE, and through F draw FG parallel to AB; and join AF: Then, because AC is equal to CE, the angle EAC is equal (I. 5.) to the angle AEC; and because the angle ACE is a right angle, the two others AEC, EAC together make one right angle (I. 32.); and they are equal to one another; each of them, therefore, is half of a right angle. For the same reason each of the angles CEB, EBC is half a right angle; and therefore the whole AEB is a right angle: And because the angle GEF is half a right angle, and EGF a right angle, for it is equal (I. 29.) to the interior and opposite angle ECB, the remaining angle EFG is half a right angle; therefore, the angle GEF is equal to the angle EFG, and the side EG equal (I. 6.) to the side GF: Again, because the angle at B is half a right angle and FDB a right angle, for it is equal (I. 29.) to the interior and opposite angle ECB, the remaining angle BFD is half a right angle; therefore the angle at B is equal to the angle BFD, and the side DF to (I. 6.) the side DB: And because AC is equal to CE, the square of AC is equal to the square of CE; therefore the squares of AC, CE are double of the square of AC: But

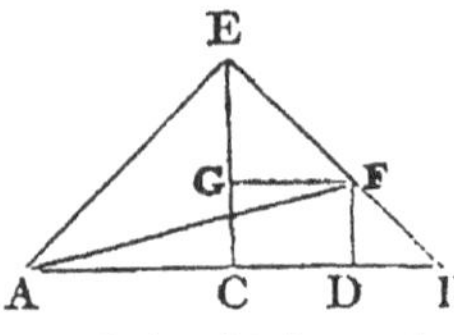

the square of EA is equal (I. 47.) to the square of AC, CE, because ACE is a right angle; therefore, the square of EA is double of the square of AC: Again, because EG is equal to GF, the square of EG is equal to the square of GF; therefore, the squares of EG, GF are double of the square of GF; but the square of EF is equal to the squares of EG, GF; therefore the square of EF is double of the square of GF; and GF is equal (I. 34.) to CD: therefore the square of EF is double of the square of CD: But the square of AE is likewise double of the square of AC: therefore the squares of AE, EF are double of the squares of AC, CD: And the square of AF is equal (I. 47.) to the squares of AE, EF, because AEF is a right angle; therefore the square of AF is double of the squares of AC, CD: But the squares of AD, DF are equal to the square of AF, because the angle ADF is a right angle; therefore the squares of AD, DF are double of the squares of AC, CD: And DF is equal to DB; therefore the squares of AD, DB are double of the squares of AC, CD. If, therefore, a straight line, &c. Q.E.D.

Prop. X. Theor.

If a straight line be bisected, and produced to any point, the square of the whole line thus produced, and the square of the part of it produced, are together double of the square of half the line bisected, and of the square of the line made up of the half and the part produced.

Let the straight line AB be bisected in C, and produced to the point D; the squares of AD, DB are double of the squares of AC, CD.

From the point C draw (I. 11.) CE at right angles to AB: And make it equal to AC or CB, and join AE, EB; through E draw (I. 31.) EF parallel to AB, and through D draw DF parallel to CE: And because the straight line EF meets the parallels EC, FD, the angles CEF, EFD are equal (I. 29.) to two right angles; and therefore the angles BEF, EFD are less than two right angles: But straight lines which with another straight line make the interior angles upon the same side less than two right angles, do meet (Ax. 12.) if produced far enough: Therefore EB, FD shall meet, if produced towards B, D: Let them meet in G, and join AG: Then, because AC is equal to CE, the angle CEA is equal (I. 5.) to the angle EAC; and the angle ACE is a right angle; therefore each of the angles CEA, EAC is half a right angle (I. 32.): For the same reason, each of the angles CEB, EBC is half a right angle; therefore AEB is a right angle: And because EBC is half a right angle, DBG is also (I. 15.) half a right angle, for they are vertically opposite; but BDG is a right angle, because it is equal (I. 29.) to the alternate angle DCE; therefore the remaining angle DGB is half a right angle, and is therefore equal to the angle DBG; wherefore also the side BD is equal (I. 6.) to the side DG: Again, because EGF is half a right angle, and that the angle at F is a right angle, because it is equal (I. 34.) to the opposite angle ECD, the remaining angle FEG is half a right angle, and equal to the angle EGF; wherefore also the side GF is equal (I. 6.) to the side FE. And because EC is equal to CA, the square of EC is equal to the square of CA; therefore the squares of EC, CA are double of the square of CA: But the square of EA is equal (I. 47.) to the squares of EC, CA; therefore the square of EA is double of the square of AC: Again, because GF is equal to FE, the square of GF is equal to the square of FE; and therefore the squares GF, FE are double of the square of EF: But the square of EG is equal (I. 47.) to the squares of GF, FE; therefore the square of EG is double of the square of EF: And EF is equal to CD; wherefore the square of EG is double of the square of CD: But it was demonstrated, that the square of EA is double of the square of AC; therefore the squares of AE, EG are double of the squares of AC, CD: And the square of AG is equal (I. 47.) to the squares of AE, EG: therefore the square of AG is double of the squares of AC, CD: But the squares of AD, DG are equal (I. 47.) to the square of AG; therefore the squares of AD, DG are double of the squares of AC, CD: But DG is equal to DB; therefore the squares of AD, DB are double of the squares of AC, CD: Wherefore, if a straight line, &c. Q.E.D.

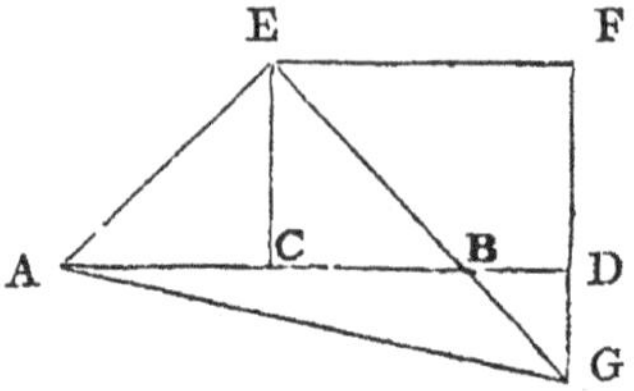

PROP. XI. PROB.

To divide a given straight line into two parts, so that the rectangle contained by the whole, and one of the parts, shall be equal to the square of the other part.

Let AB be the given straight line; it is required to divide it into two parts, so that the rectangle contained by the whole, and one of the parts, shall be equal to the square of the other part.

Upon AB describe (I. 46.) the square ABDC; bisect (I. 10.) AC in E, and join BE; produce CA to F, and make (I. 3.) EF equal to EB, and upon AF describe (I. 46.) the square FGHA; AB is divided in H, so that the rectangle AB, BH is equal to the square of AH.

Produce GH to K: Because the straight line AC is bisected in E, and produced to the point F, the rectangle CF, FA, together with the square of AE, is equal (II. 6.) to the square of EF: But EF is equal to EB; therefore the rectangle CF, FA, together with the square of AE, is equal to the square of EB: And the squares of BA, AE are equal (I. 47.) to the square of EB, because the angle EAB is a right angle; therefore the rectangle CF, FA together with the square of AE, is equal to the squares of BA, AE: Take away the square of AE, which is common to both, therefore the remaining rectangle CF, FA is equal to the square of AB: and the figure FK is the rectangle contained by CF, FA, for AF is equal to FG; and AD is the square of AB; therefore FK is equal to AD: Take away the common part AK, and the remainder FH is equal to the remainder HD: And HD is the rectangle contained by AB, BH, for AB is equal to BD; and FH is the square of AH. Therefore the rectangle AB, BH is equal to the square of AH: Wherefore the straight line AB is divided in H, so that the rectangle AB, BH is equal to the square of AH. Which was to be done.

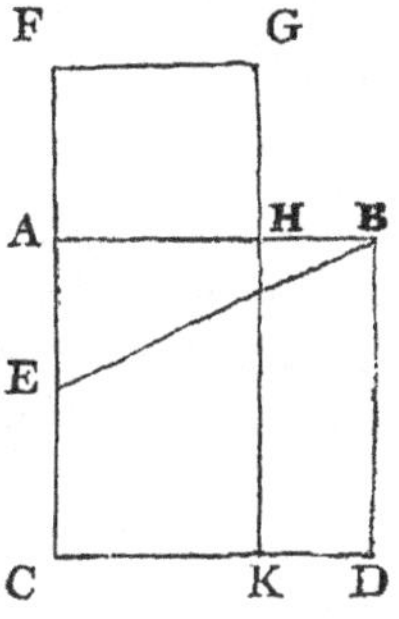

Prop. XII. Theor.

In obtuse angled triangles, if a perpendicular be drawn from any of the acute angles to the opposite side produced, the square of the side subtending the obtuse angle is greater than the squares of the sides containing the obtuse angle, by twice the rectangle contained by the side upon which, when produced, the perpendicular falls, and the straight line intercepted without the triangle between the perpendicular and the obtuse angle.

Let ABC be an obtuse angled triangle, having the obtuse angle ACB, and from the point A let AD be drawn (I. 12.) perpendicular to BC produced: the square of AB is greater than the squares of AC, CB by twice the rectangle BC, CD.

Because the straight line BD is divided into two parts in the point C, the square of BD is equal (II. 4.) to the squares of BC, CD, and twice the rectangle BC, CD: To each of these equals add the square of DA; and the squares of DB, DA are equal to the squares of BC, CD, DA, and twice the rectangle BC, CD: But the square of BA is equal (I. 47.) to the squares of BD, DA, because the angle at D is a right angle; and the square of CA is equal (I. 47.) to the squares of CD, DA: Therefore the square of BA is equal to the squares of BC, CA, and twice the rectangle BC, CD; that is, the square of BA is greater than the squares of BC, CA, by twice the rectangle BC, CD. Therefore, in obtuse angled triangles, &c. Q. E. D.

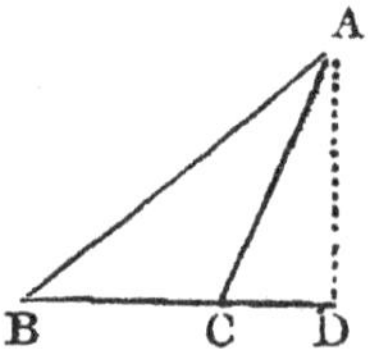

Prop. XIII. Theor.

In every triangle, the square of the side subtending any of the acute angles is less than the squares of the sides containing that angle, by twice the rectangle contained by either of these sides, and the straight line intercepted between the perpendicular let fall upon it from the opposite angle, and the acute angle.

Let ABC be any triangle, and the angle at B one of its acute angles, and upon BC, one of the sides containing it, let fall the perpendicular (I. 12.) AD from the opposite angle:

The square of AC opposite to the angle B, is less than the squares of CB, BA by twice the rectangle CB, BD.

First, let AD fall within the triangle ABC; and because the straight line CB is divided into two parts in the point D, the squares of CB, BD are equal (II. 7.) to twice the rectangle contained by CB, BD, and the square of DC: To each of these equals add the square of AD; therefore the squares of CB, BD, DA are equal to twice the rectangle CB, BD, and the squares of AD, DC: But the square of AB is equal (I. 47.) to the squares of BD, DA, because the angle BDA is a right angle; and the square of AC is equal to the squares of AD, DC: Therefore the squares of CB, BA are equal to the square of AC, and twice the rectangle CB, BD, that is, the square of AC alone is less than the squares of CB, BA by twice the rectangle CB, BD.

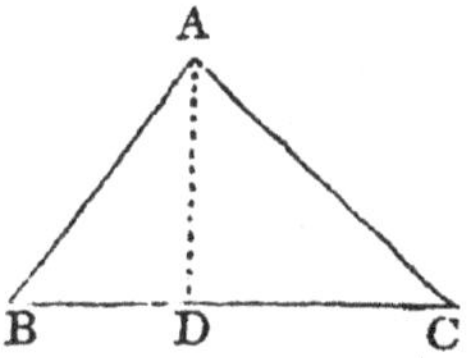

Secondly, let AD fall without the triangle ABC: Then, because the angle at D is a right angle, the angle ACB is greater (I. 16.) than a right angle; and therefore the square of AB is equal (II. 12.) to the squares of AC, CB, and twice the rectangle BC, CD: To these equals add the square of BC, and the squares of AB, BC are equal to the square of AC, and twice the square of BC, and twice the rectangle BC, CD: But because BD is divided into two parts in C, the rectangle DB, BC is equal (II. 3.) to the rectangle BC, CD and the square of BC: And the doubles of these are equal: Therefore the squares of AB, BC are equal to the square of AC, and twice the rectangle DB, BC: Therefore the square of AC alone is less than the squares of AB, BC by twice the rectangle DB, BC.

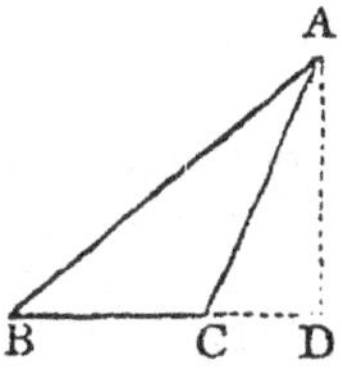

Lastly, let the side AC be perpendicular to BC; then is BC the straight line between the perpendicular and the acute angle at B; and it is manifest that the squares of AB, BC are equal (I. 47.) to the square of AC, and twice the square of BC: Therefore in every triangle, &c. Q. E. D.

PROP. XIV. PROB.

To describe a square that shall be equal to a given rectilineal figure.

Let A be the given rectilineal figure; it is required to describe a square that shall be equal to A.

Describe (I. 45.) the rectangular parallelogram BCDE equal to the rectilineal figure A. If then the sides of it BE, ED are equal to one another, it is a square, and what was required is now done: But if they are not equal, produce one of them BE to F, and make EF equal to ED, and bisect BF in G; and from the centre G, at the distance GB, or GF, describe the semicircle BHF, and produce DE to H, and join GH.

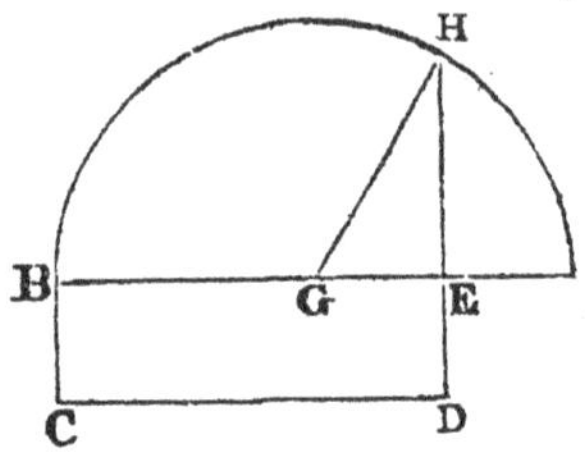

Then, because the straight line BF is divided into two equal parts in the point G, and into two unequal at E, the rectangle BE, EF, together with the square of EG, is equal (II. 5.) to the square of GF: But GF is equal to GH; therefore the rectangle BE, EF, together with the square of EG, is equal to the square of GH: But the squares of HE, EG are equal (I. 47.) to the square of GH: Therefore the rectangle BE, EF, together with the square of EG, is equal to the squares of HE, EG: Take away the square of EG, which is common to both, and the remaining rectangle BE, EF is equal to the square of EH: But the rectangle contained by BE, EF is the parallelogram BD, because EF is equal to ED; therefore BD is equal to the square of EH; but BD is equal to the rectilineal figure A; therefore the rectilineal figure A is equal to the square of EH: Wherefore a square has been made equal to the given rectilineal figure A, viz. the square described upon EH. Which was to be done.

EUCLID'S

ELEMENTS OF GEOMETRY.

BOOK III.

DEFINITIONS.

I.

Equal circles are those of which the diameters are equal, or from the centres of which the straight lines to the circumferences are equal.

"This is not a definition but a theorem, the truth of which is evident; for, if the circles be applied to one another, so that their centres coincide, the circles must likewise coincide, since the straight lines from the centres are equal."

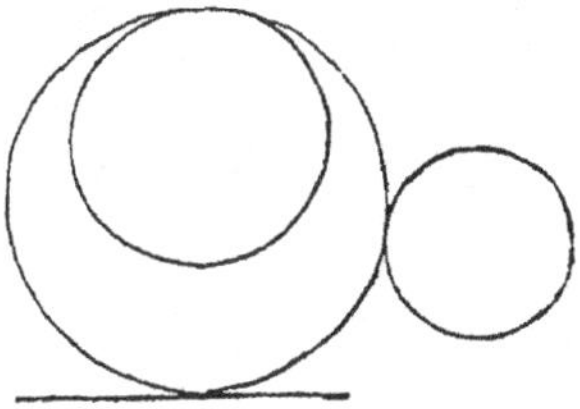

II.

A straight line is said to touch a circle, when it meets the circle, and being produced does not cut it.

III.

Circles are said to touch one another, which meet, but do not cut one another.

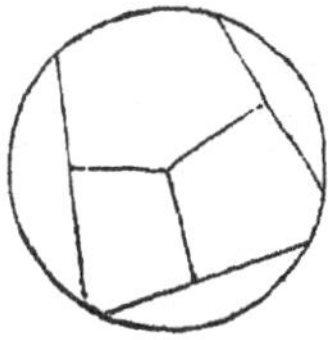

IV.

Straight lines are said to be equally distant from the centre of a circle, when the perpendiculars drawn to them from the centre are equal.

V.

And the straight line on which the greater perpendicular falls is said to be farther from the centre.

VI.

A segment of a circle is the figure contained by a straight line and the circumference it cuts off.

VII.

"The angle of a segment is that which is contained by the straight line and the circumference."

VIII.

An angle in a segment is the angle contained by two straight lines drawn from any point in the circumference of the segment, to the extremities of the straight line which is the base of the segment.

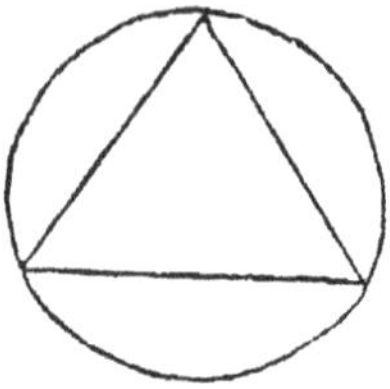

IX.

And an angle is said to insist or stand upon the circumference intercepted between the straight lines that contain the angle.

X.

The sector of a circle is the figure contained by two straight lines drawn from the centre, and the circumference between them.

XI.

Similar segments of a circle are those in which the angles are equal, or which contain equal angles.

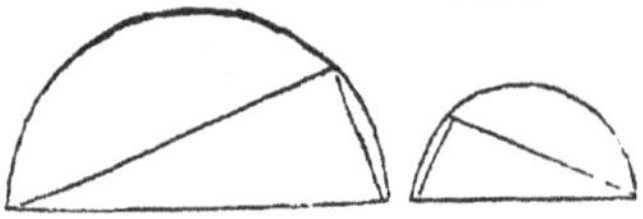

PROP. I. PROB.

To find the centre of a given circle.

Let ABC be the given circle; it is required to find its centre.

Draw within it any straight line AB, and bisect (I. 10.) it in D; from the point D draw (I. 11.) DC at right angles to AB, and produce it to E, and bisect CE in F: The point F is the centre of the circle ABC.

For, if the centre be in CE, it must be in the middle point of CE, that is, it must be in F:

But if it be not in CE, let, if possible, G be the centre, and join GA, GD, GB: Then, because DA is equal to DB, and DG common to the two triangles ADG, BDG, the two sides AD, DG are equal to the two BD, DG, each to each; and the base GA is equal to the base GB, because they are drawn from the centre G*; therefore the angle ADG is equal (I. 8.) to the angle GDB: But when a straight line standing upon another straight line makes the adjacent angles equal to one another, each of the angles is a right angle (I. Def. 10.): Therefore the angle GDB is a right angle: But FDB is likewise a right angle; wherefore the angle FDB is equal to the angle GDB, the greater to the less, which is impossible: Therefore G is not the centre of the circle ABC: In the same manner it can be shown, that no other point but F is the centre; that is, F is the centre of the circle ABC: Which was to be found.

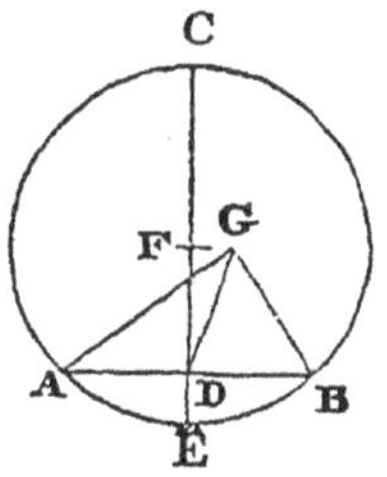

COR. From this it is manifest, that if in a circle a straight line bisect another at right angles, the centre of the circle is in the line which bisects the other.

PROP. II. THEOR.

If any two points be taken in the circumference of a circle, the straight line which joins them shall fall within the circle.

Let ABC be a circle, and A, B any two points in the circumference; the straight line drawn from A to B shall fall within the circle.

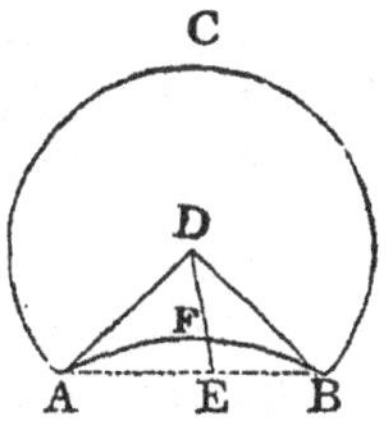

For, if it do not, let it fall, if possible, without, as AEB; find (III. 1.) D, the centre of the circle ABC, and join AD, DB; in the circumference AB take any point F, join DF, and produce it to E: Then, because DA is equal to DB, the angle DAB is equal (I. 5.) to the angle DBA; and because AE, a side of the triangle DAE, is produced to B, the angle DEB is greater (I. 16.) than the angle DAE; but DAE is equal to the angle DBE; therefore the angle DEB is greater than the angle DBE; but to the

* N. B. — Whenever the expression "straight lines from the centre," or "drawn from the centre," occurs, it is to be understood that they are drawn to the circumference.

greater angle the greater side is opposite (I. 19.); DB is therefore greater than DE: But DB is equal to DF; wherefore DF is greater than DE, the less than the greater, which is impossible: Therefore the straight line drawn from A to B does not fall without the circle. In the same manner it may be demonstrated that it does not fall upon the circumference; it falls therefore within it. Wherefore if any two points, &c. Q.E.D.

PROP. III. THEOR.

If a straight line drawn through the centre of a circle bisect a straight line in it which does not pass through the centre, it shall cut it at right angles; and, if it cuts it at right angles, it shall bisect it.

Let ABC be a circle, and let CD, a straight line drawn through the centre, bisect any straight line AB which does not pass through the centre, in the point F: It cuts it also at right angles.

Take (III. 1.) E the centre of the circle, and join EA, EB. Then, because AF is equal to FB, and FE common to the two triangles AFE, BFE, there are two sides in the one equal to two sides in the other, and the base EA is equal to the base EB; therefore the angle AFE is equal (I. 8.) to the angle BFE: But when a straight line standing upon another makes the adjacent angles equal to one another, each of them is a right (I. Def. 10.) angle: Therefore each of the angles AFE, BFE is a right angle; wherefore the straight line CD, drawn through the centre bisecting another AB that does not pass through the centre, cuts the same at right angles.

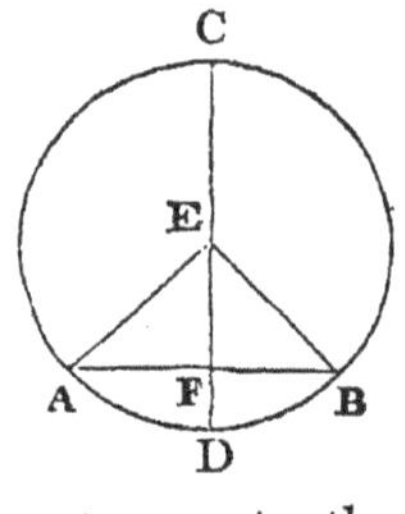

But let CD cut AB at right angles; CD also bisects it, that is, AF is equal to FB.

The same construction being made, because EA, EB from the centre are equal to one another, the angle EAF is equal (I. 5.) to the angle EBF; and the right angle AFE is equal to the right angle BFE: Therefore in the two triangles EAF, EBF, there are two angles in one equal to two angles in the other, and the side EF, which is opposite to one of the equal angles in each, is common to both; therefore the other sides are equal (I. 26.); AF therefore is equal to FB. Wherefore if a straight line, &c. Q.E.D

Prop. IV. Theor.

If in a circle two straight lines cut one another which do not both pass through the centre, they do not bisect each the other.

Let ABCD be a circle, and AC, BD two straight lines in it which cut one another in the point E, and do not both pass through the centre; AC, BD do not bisect one another.

For, if it is possible, let AE be equal to EC, and BE to ED: If one of the lines pass through the centre, it is plain that it cannot be bisected by the other which does not pass through the centre: But if neither of them pass through the centre, take (III. 1.) F the centre of the circle, and join EF: and because FE, a straight line through the centre, bisects another AC which does not pass through the centre, it shall cut it at right (III. 3.) angles; wherefore FEA is a right angle: Again, because the straight line FE bisects the straight line BD which does not pass through the centre, it shall cut it at right (III. 3.) angles; wherefore FEB is a right angle: And FEA was shown to be a right angle; therefore FEA is equal to the angle FEB, the less to the greater, which is impossible: Therefore AC, BD do not bisect one another. Wherefore, if in a circle, &c. Q.E.D.

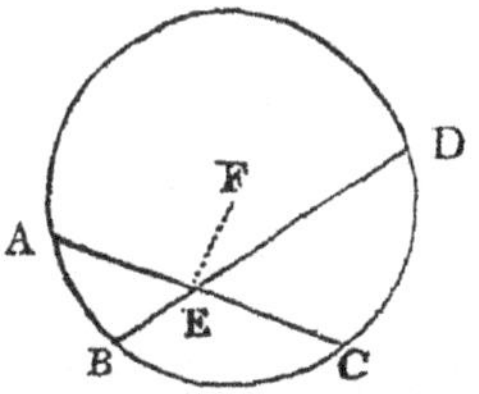

Prop. V. Theor.

If two circles cut one another, they shall not have the same centre.

Let the two circles ABC, CDG cut one another in the points B, C; they have not the same centre.

For, if it be possible, let E be their centre: Join EC, and draw any straight line EFG meeting them in F and G: And because E is the centre of the circle ABC, CE is equal to EF: Again, because E is the centre of the circle CDG, CE is equal to EG: But CE was shown to be equal to EF, therefore EF is equal to EG, the less to the greater, which is impossible: Therefore E is not the

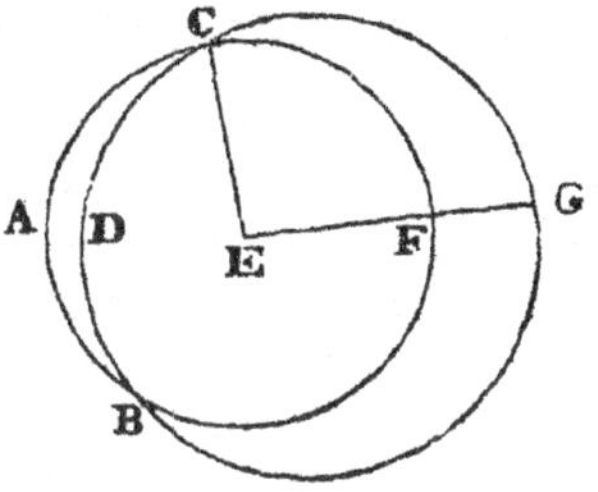

centre of the circles ABC, CDG. Wherefore if two circles, &c. Q.E.D.

PROP. VI. THEOR.

If two circles touch one another internally, they shall not have the same centre.

Let the two circles ABC, CDE, touch one another internally in the point C: They have not the same centre.

For, if they can, let it be F; join FC and draw any straight line FEB meeting them in E and B; and because F is the centre of the circle ABC, CF is equal to FB: Also because F is the centre of the circle CDE, CF is equal to FE. And CF was shown equal to FB; therefore FE is equal to FB, the less to the greater, which is impossible; wherefore F is not the centre of the circles ABC, CDE. Therefore if two circles, &c Q.E.D.

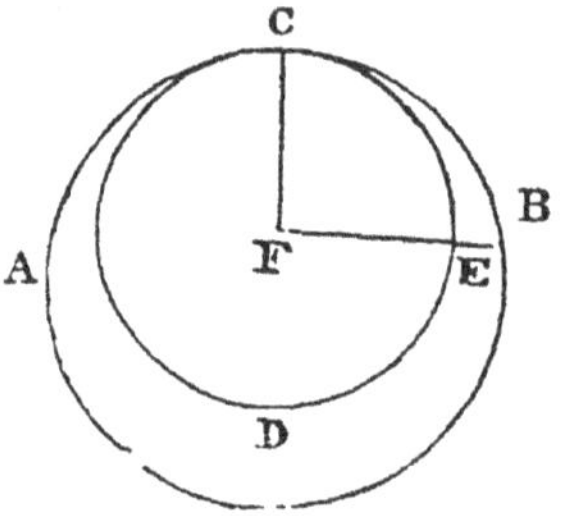

PROP. VII. THEOR.

If any point be taken in the diameter of a circle which is not the centre, of all the straight lines which can be drawn from it to the circumference, the greatest is that in which the centre is, and the other part of that diameter is the least; and, of any others, that which is nearer to the line which passes through the centre is always greater than one more remote: and from the same point there can be drawn only two straight lines that are equal to one another, one upon each side of the shortest line.

Let ABCD be a circle, and AD its diameter, in which let any point F be taken which is not the centre: Let the centre be E; of all the straight lines FB, FC, FG, &c., that can be drawn from F to the circumference, FA is the greatest, and FD, the other part of the diameter AD, is the least: and of the others, FB is greater than FC, and FC than FG.

Join BE, CE, GE; and because two sides of a triangle are

greater (I. 20.) than the third, BE, EF are greater than BF; but AE is equal to EB; therefore AE, EF, that is AF, is greater than BF: Again, because BE is equal to CE, and FE common to the triangles BEF, CEF, the two sides BE, EF are equal to the two CE, EF; but the angle BEF is greater than the angle CEF; therefore the base BF is greater (I. 24.) than the base FC: For the same reason, CF is greater than GF: Again, because GF, FE are greater (I. 20.) than EG, and EG is equal to ED; therefore GF, FE are greater than ED: Take away the common part FE, and the remainder GF is greater than the remainder FD: Therefore FA is the greatest, and FD the least of all the straight lines from F to the circumference; and BF is greater than CF, and CF than GF.

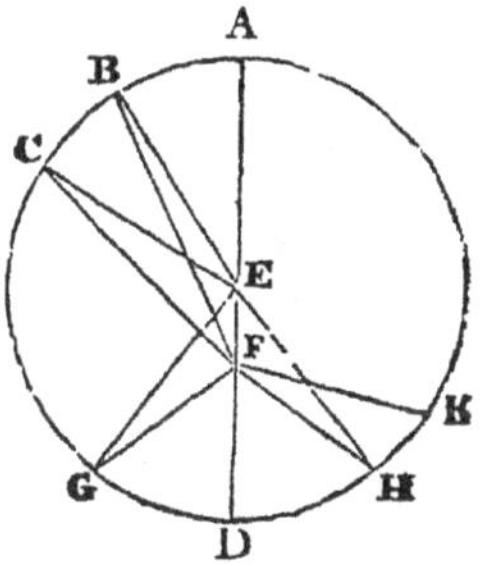

Also there can be drawn only two equal straight lines from the point F to the circumference, one upon each side of the shortest line FD: At the point E in the straight line EF, make (I. 23.) the angle FEH equal to the angle GEF, and join FH: Then because GE is equal to EH, and EF common to the two triangles GEF, HEF; the two sides GE, EF are equal to the two HE, EF; and the angle GEF is equal to the angle HEF; therefore the base FG is equal (I. 4.) to the base FH: But, besides FH, no other straight line can be drawn from F to the circumference equal to FG: For, if there can, let it be FK; and because FK is equal to FG, and FG to FH, FK is equal to FH; that is, a line nearer to that which passes through the centre, is equal to that which is more remote; which is impossible. Therefore, if any point be taken, &c. Q. E. D.

PROP. VIII. THEOR.

If any point be taken without a circle, and straight lines be drawn from it to the circumference, whereof one passes through the centre; of those which fall upon the concave circumference, the greatest is that which passes through the centre; and of the rest, that which is nearer to the one passing through the centre is always greater than one more remote: But of those which fall upon the convex circumference, the least is that between the point without the circle and the diameter; and of the rest, that which is nearer to the least is always less than one more remote: And only two equal straight lines can be drawn from the same point to the circumference, one upon each side of the least line.

Let ABC be a circle, and D any point without it, from which let the straight lines DA, DE, DF, DC be drawn to the circumference, whereof DA passes through the centre. Of those which fall upon the concave part of the circumference AEFC, the greatest is AD which passes through the centre; and the nearer to it is always greater than the more remote, viz. DE than DF, and DF than DC: But of those which fall upon the convex circumference HLKG, the least is DG betwen the point D and the diameter AG; and the nearer to it is always less than the more remote, viz. DK than DL, and DL than DH.

Take (III. 1.) M the centre of the circle ABC, and join ME, MF, MC, MK, ML, MH: And because AM is equal to ME, add MD to each, therefore AD is equal to EM, MD; but EM, MD are greater (I. 20.) than ED; therefore also AD is greater than ED: Again, because ME is equal to MF, and MD common to the triangles EMD, FMD; EM, MD are equal to FM, MD; but the angle EMD is greater than the angle FMD; therefore the base ED is greater (I. 24.) than the base FD: In like manner it may be shown that FD is greater than CD: Therefore DA is the greatest; and DE greater than DF, and DF than DC: And because MK, KD are greater than MD, and MK is equal to MG, the remainder KD is greater (Ax. 5.) than the remainder GD, that is, GD is less than KD: And because MK, DK are drawn to the point K within the triangle MLD, from M, D, the extremities of its side MD, therefore MK, KD are less (I. 21.) than ML, LD, whereof MK is equal to ML; therefore the remainder DK is less than the remainder DL: In like manner it may be shown, that DL is less than DH: Therefore DG is the least, and DK less than DL, and DL than DH: Also there can be drawn only two equal straight lines from the point D to the circumference,

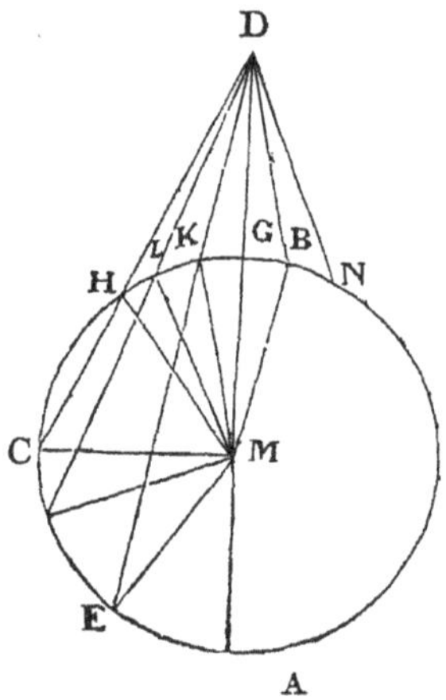

one upon each side of the least: At the point M, in the straight line MD, make (I. 23.) the angle DMB equal to the angle DMK, and join DB: And because MK is equal to MB, and MD common to the triangles KMD, BMD, the two sides KM, MD are equal to the two BM, MD; and the angle KMD is equal to the angle BMD; therefore the base DK is equal (I. 4.) to the base DB: But, besides DB, there can be no straight line drawn from D to the circumference equal to DK: For, if there can, let it be DN; and because DK is equal to DN, and also to DB; therefore DB is equal to DN, that is, the nearer to the least equal to the more remote, which is impossible. If, therefore, any point, &c. Q. E. D.

Prop. IX. Prob.

If a point be taken within a circle, from which there fall more than two equal straight lines to the circumference, that point is the centre of the circle.

Let the point D be taken within the circle ABC, from which to the circumference there fall more than two equal straight lines, viz. DA, DB, DC, the point D is the centre of the circle.

For, if not, let E be the centre, join DE and produce it to the circumference in F, G; then FG is a diameter of the circle ABC: And because in FG, the diameter of the circle ABC, there is taken the point D which is not the centre, DG shall be the greatest line from it to the circumference, and DC greater (III. 7.) than DB, and DB than DA: But they are likewise equal (Hyp.), which is impossible: Therefore E is not the centre of the circle ABC: In like manner it may be demonstrated that no other point but D is the centre; D therefore is the centre. Wherefore, if a point be taken, &c. Q. E. D.

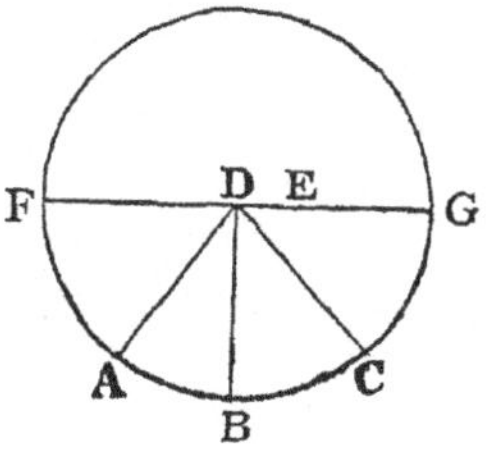

Prop. X. Theor.

One circumference of a circle cannot cut another in more than two points.

If it be possible, let the circumference FAB cut the circumference DEF in more than two points, viz. in B, G, F; take the centre K of the circle ABC, and join KB, KG, KF: And because within the circle DEF there is taken the point K, from which to the circumference DEF fall more than two equal straight lines KB, KG, KF, the point K is (III. 9.) the centre of the circle DEF: But K is also the centre of the circle ABC: therefore the same point is the centre of two circles that cut one another, which is impossible (III. 5.). Therefore one circumference of a circle cannot cut another in more than two points. Q. E. D.

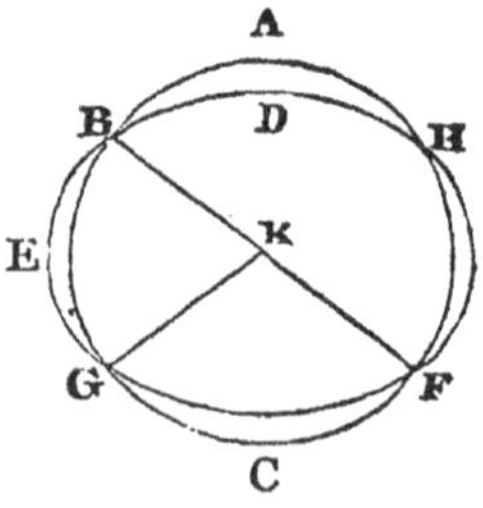

PROP. XI. THEOR.

If two circles touch each other internally, the straight line which joins their centres being produced shall pass through the point of contact.

Let the two circles ABC, ADE, touch each other internally in the point A, and let F be the centre of the circle ABC, and G the centre of the circle ADE: The straight line which joins the centres F, G, being produced, passes through the point A.

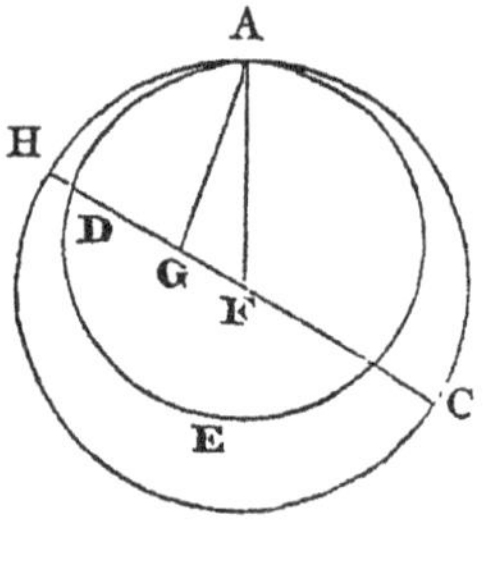

For, if not, let it fall otherwise, if possible, as FGDH, and join AF, AG: And because AG, GF are greater (I. 20.) than FA, that is, than FH, for FA is equal to FH, both being from the same centre; take away the common part FG; therefore the remainder AG is greater than the remainder GH: But AG is equal to GD; therefore GD is greater than GH, the less than the greater, which is impossible. Therefore the straight line which joins the points F, G cannot fall otherwise than upon the point A, that is, it must pass through it. Therefore, if two circles, &c. Q. E. D.

PROP. XII. THEOR.

If two circles touch each other externally, the straight line which joins their centres shall pass through the point of contact.

Let the two circles ABC, ADE touch each other externally in the point A: and let F be the centre of the circle ABC, and G the centre of ADE: The straight line which joins the points F, G shall pass through the point of contact A.

For, if not, let it pass otherwise, if possible, as FCDG, and join FA, AG: And because F is the centre of the circle ABC, AF is equal to FC: Also because G is the centre of the circle ADE, AG is equal to GD: Therefore FA, AG are equal to FC, DG; wherefore the whole FG is greater than FA, AG; But it is also less (I. 20.); which is impossible: Therefore the straight line which joins the points F, G shall not pass otherwise than through the point of contact A; that is, it must pass through it. Therefore, if two circles, &c. Q. E. D.

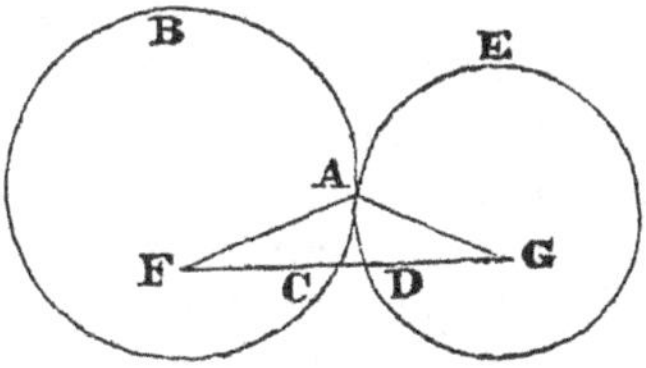

PROP. XIII. THEOR.

One circle cannot touch another in more points than one, whether it touches it on the inside or outside.

For, if it be possible, let the circle EBF touch the circle ABC in more points than one, and first on the inside, in the points B, D; join BD, and draw (I. 10, 11.) GH bisecting BD at right angles: Therefore, because the points B, D are in

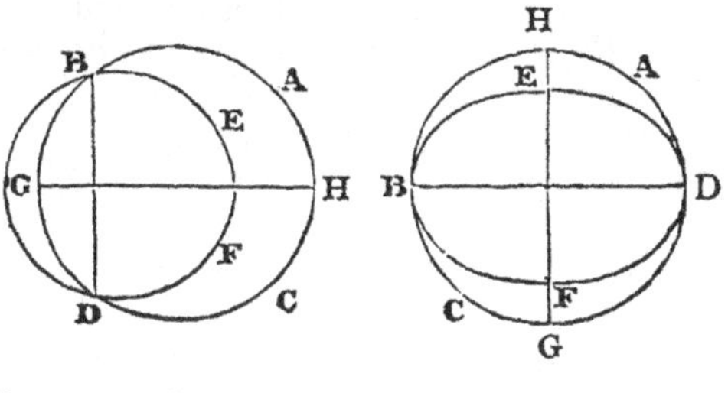

the circumference of each of the circles, the straight line BD falls within each (III. 2.) of them: And their centres are

(III. 1. Cor.) in the straight line GH which bisects BD at right angles: Therefore GH passes through the point of contact (III. 11.); but it does not pass through it, because the points B, D are without the straight line GH, which is absurd: Therefore one circle cannot touch another on the inside in more points than one.

Nor can two circles touch one another on the outside in more than one point: For, if it be possible, let the circle ACK touch the circle ABC in the points A, C, and join AC: Therefore, because the two points A, C are in the circumference of the circle ACK, the straight line AC which joins them shall fall within (III. 2.) the circle ACK: And the circle ACK is without the circle ABC; and therefore the straight line AC is without this last circle; but, because the points A, C are in the circumference of the circle ABC, the straight line AC must be within (III. 2.) the same circle, which is absurd: Therefore one circle cannot touch another on the outside in more than one point: And it has been shown, that they cannot touch on the inside in more points than one: Therefore, one circle, &c. Q. E. D.

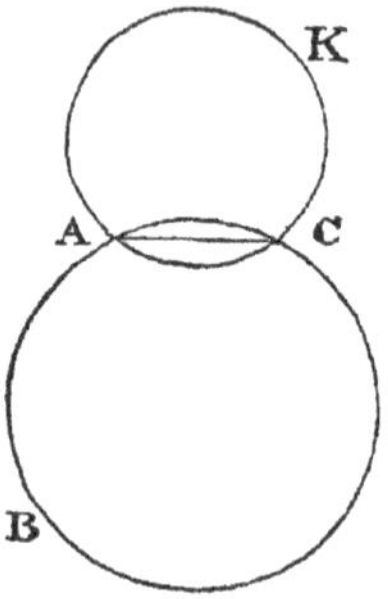

PROP. XIV. THEOR.

Equal straight lines in a circle are equally distant from the centre; and those which are equally distant from the centre, are equal to one another.

Let the straight lines AB, CD, in the circle ABDC, be equal to one another; they are equally distant from the centre.

Take E the centre of the circle ABDC, and from it draw EF, EG perpendiculars to AB, CD: Then, because the straight line EF passing through the centre, cuts the straight line AB, which does not pass through the centre, at right angles, it also bisects (III. 3.) it: Wherefore AF is equal to FB, and AB double of AF. For the same reason, CD is double of CG: And AB is equal to CD (Hyp.); therefore, AF is equal to CG: And because AE is equal to EC, the square of AE is equal to the square of EC: But the squares of AF, FE are equal (I. 47.) to

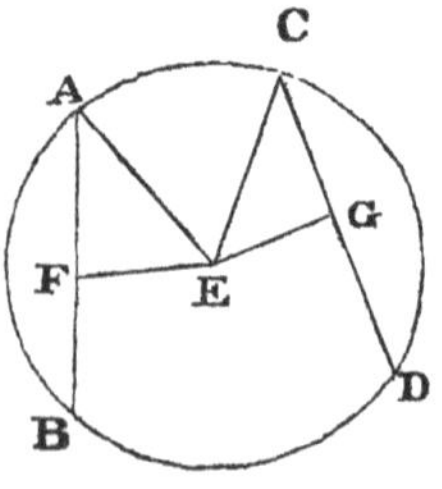

the square of AE, because the angle AFE is a right angle: and, for the like reason, the squares of EG, GC are equal to the square of EC: Therefore the squares of AF, FE are equal to the squares of CG, GE, of which the square of AF is equal to the square of CG, because AF is equal to CG; therefore the remaining square of FE is equal to the remaining square of EG, and the straight line EF is therefore equal to EG: but straight lines in a circle are said to be equally distant from the centre, when the perpendiculars drawn to them from the centre are equal (III. Def. 4.): Therefore AB, CD are equally distant from the centre.

Next, if the straight lines AB, CD be equally distant from the centre, that is, if FE be equal to EG; AB is equal to CD: For, the same construction being made, it may, as before, be demonstrated, that AB is double of AF, and CD double of CG, and that the squares of EF, FA are equal to the squares of EG, GC; of which the square of FE is equal to the square of EG, because FE is equal to EG; therefore the remaining square of AF is equal to the remaining square of CG; and the straight line AF is therefore equal to CG: But AB is double of AF, and CD double of CG; wherefore AB is equal to CD. Therefore equal straight lines, &c. Q. E. D.

Prop. XV. Theor.

The diameter is the greatest straight line in a circle; and, of all others, that which is nearer to the centre is always greater than one more remote; and the greater is nearer to the centre than the less.

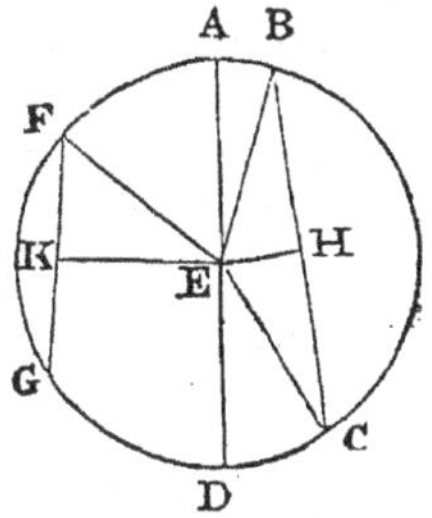

Let ABCD be a circle, of which the diameter is AD, and the centre E; and let BC be nearer to the centre than FG; AD is greater than any straight line BC which is not a diameter, and BC greater than FG.

From the centre draw EH, EK perpendiculars to BC, FG, and join EB, EC, EF; and because AE is equal to EB, and ED to EC: AD is equal to EB, EC: but EB, EC, are greater (I. 20.) than BC; wherefore also AD is greater than BC.

And because BC is nearer to the centre than FG, EH is less (III. Def. 5.) than EK: But, as was demonstrated in the preceding, BC is double of BH, and FG double of FK, and the squares of EH, BH are equal to the squares of EK, KF, of which the square of EH is less than the square of EK, because EH is less than EK; therefore the square of BH is greater than the square of FK, and the straight line BH greater than FK; and therefore BC is greater than FG.

Next, let BC be greater than FG; BC is nearer to the centre than FG, that is, the same construction being made, EH is less than EK: because BC is greater than FG, BH likewise is greater than KF: And the squares of BH, HE are equal to the squares of FK, KE, of which the square of BH is greater than the square of FK, because BH is greater than FK; therefore the square of EH is less than the square of EK, and the straight line EH less than EK. Wherefore the diameter, &c. Q. E. D.

Prop. XVI. Theor.

The straight line drawn at right angles to the diameter of a circle, from the extremity of it, falls without the circle; and no straight line can be drawn between that straight line and the circumference from the extremity, so as not to cut the circle; or, which is the same thing, no straight line can make so great an acute angle with the diameter at its extremity, or so small an angle with the straight line which is at right angles to it, as not to cut the circle.

Let ABC be a circle, the centre of which is D, and the diameter AB: the straight line drawn at right angles to AB from its extremity A, shall fall without the circle.

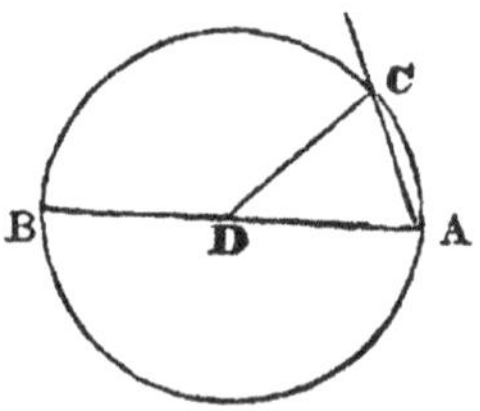

For, if it does not, let it fall, if possible, within the circle, as AC, and draw DC to the point C where it meets the circumference: And because DA is equal to DC, the angle DAC is equal (I. 5.) to the angle ACD; but DAC is a right angle, therefore ACD is a right angle, and the angles DAC, ACD are therefore equal to two right angles; which is impossible (I. 17.): Therefore the straight line drawn from A at right angles to BA does not fall within the circle: In the same manner, it may be demonstrated that it does not fall upon the circumference: therefore it must fall without the circle, as AE.

And between the straight line AE and the circumference no straight line can be drawn from the point A which does not cut the circle: For, if possible, let FA be between them, and from the point D draw (I. 12.) DG perpendicular to FA, and let it meet the circumference in H: And because AGD is a right angle, and DAG less (I. 17.) than a right angle: DA is greater (I. 19.), than DG: But DA is equal to DH; therefore DH is greater than DG, the less than the greater, which is impossible: Therefore no straight line can be drawn from the point A between AE and the circumference, which does not cut the circle, or, which amounts to the same thing, however great an acute angle a straight line makes with the diameter at the point A, or however small an angle it makes with AE, the circumference passes between the straight line and the perpendicular AE. "And this is all that is to be understood, when, in the Greek text and translations from it, the angle of the semicircle is said to be greater than any acute rectilineal angle, and the remaining angle less than any rectilineal angle."

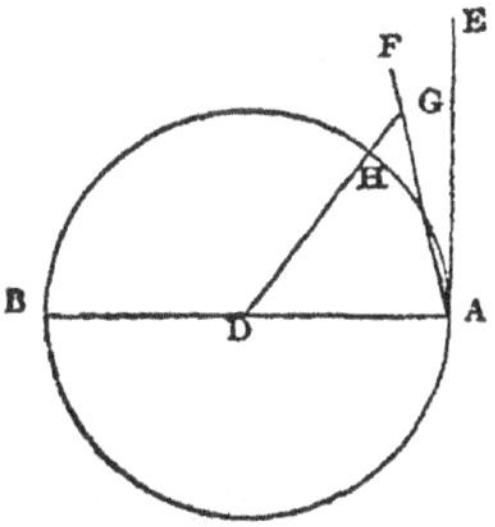

COR. From this it is manifest that the straight line which is drawn at right angles to the diameter of a circle from the extremity of it, touches the circle; and that it touches it only in one point, because, if it did meet the circle in two, it would fall within it (III. 2.). "Also it is evident that there can be but one straight line which touches the circle in the same point."

PROP. XVII. PROB.

To draw a straight line from a given point, either without or in the circumference, which shall touch a given circle.

First, let A be a given point without the given circle BCD, it is required to draw a straight line from A which shall touch the circle.

Find (III. 1.) the centre E of the circle, and join AE; and from the centre E, at the distance AE, describe the circle AFG; from the point D draw (I. 11.) FD at right angles to AE; and join EBF, AB. AB touches the circle BCD.

Because E is the centre of the circles BCD, AFG, EA is equal to EF, and ED to EB; therefore the two sides AE, EB are equal to the two FE, ED, and they contain the angle at E common to the two triangles AEB, FED; therefore the base DF is equal to the base AB, and the triangle FED to the triangle AEB, and the other angles to the other angles (I. 4.): Therefore the angle EBA is equal to the angle EDF: But EDF is a right angle, wherefore EBA is a right angle: And EB is drawn from the centre: but a straight line drawn from the extremity of a diameter, at right angles to it, touches the circle (III. 16. Cor.): Therefore AB touches the circle; and it is drawn from the given point A. Which was to be done.

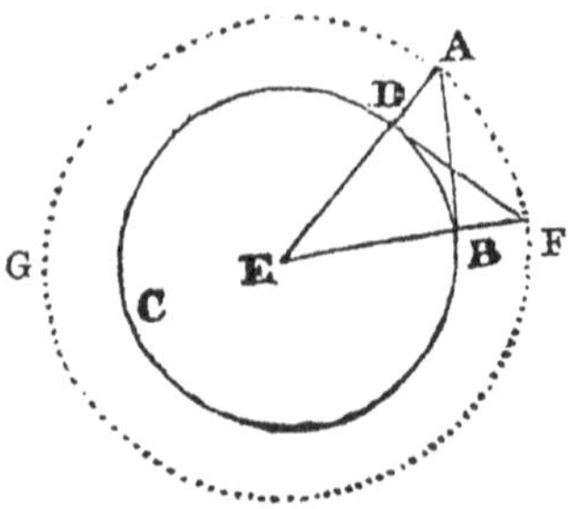

But, if the given point be in the circumference of the circle, as the point D, draw DE to the centre E and DF at right angles to DE; DF touches the circle (III. 16. Cor.).

Prop. XVIII. Theor.

If a straight line touches a circle, the straight line drawn from the centre to the point of contact, shall be perpendicular to the line touching the circle.

Let the straight line DE touch the circle ABC in the point C; take the centre F, and draw the straight line FC; FC is perpendicular to DE.

For, if it be not, from the point F draw FBG perpendicular to DE; and because FGC is a right angle, GCF is (I. 17.) an acute angle; and to the greater angle the greatest (I. 19.) side is opposite; therefore FC is greater than FG; but FC is equal to FB; therefore FB is greater than FG, the less than the greater which is impossible: Wherefore FG is not perpendicular to DE: In the same manner it may be shown, that no other is perpendicular to it besides FC, that is, FC is perpendicular to DE. Therefore, if a straight line, &c. Q. E. D.

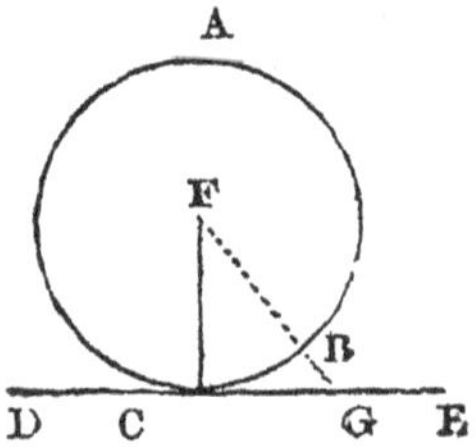

Prop. XIX. Theor.

If a straight line touches a circle, and from the point of contact a straight line be drawn at right angles to the touching line, the centre of the circle shall be in that line.

Let the straight line DE touch the circle ABC in C, and from C let CA be drawn at right angles to DE; the centre of the circle is in CA.

For, if not, let F be the centre, if possible, and join CF. Because DE touches the circle ABC, and FC is drawn from the centre to the point of contact, FC is perpendicular (III. 18.) to DE; therefore FCE is a right angle: But ACE is also a right angle; therefore the angle FCE is equal to the angle ACE, the less to the greater, which is impossible: Wherefore F is not the centre of the circle ABC. In the same manner, it may be shown, that no other point which is not in CA, is the centre; that is, the centre is in CA. Therefore if a straight line, &c. Q.E.D.

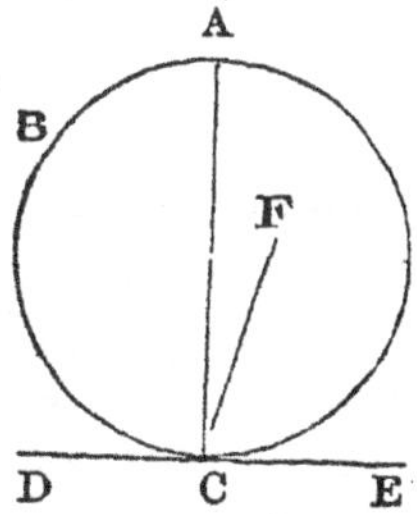

Prop. XX. Theor.

The angle at the centre of the circle is double of the angle at the circumference, upon the same base, that is, upon the same part of the circumference.

Let ABC be a circle, and BEC an angle at the centre, and BAC an angle at the circumference, which have the same circumference BC for their base; the angle BEC is double of the angle BAC.

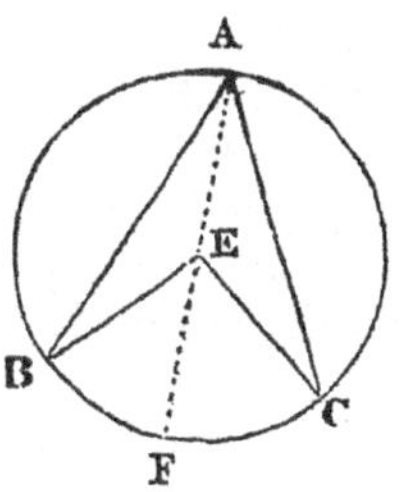

First, let E the centre of the circle be within the angle BAC, and join AE, and produce it to F: Because EA is equal to EB, the angle EAB is equal (I. 5.) to the angle EBA; therefore the angles EAB, EBA are double of the angle EAB; but the angle BEF is equal (I. 32.) to the angles EAB, EBA; therefore also the angle BEF is double of the angle EAB: For the same reason, the angle FEC is double of the angle EAC: Therefore the whole angle BEC is double of the whole angle BAC.

Again, let E the centre of the circle be without the angle BDC, and join DE and produce it to G. It may be demonstrated, as in the first case, that the angle GEC is double of the angle GDC, and that GEB a part of the first is double of GDB a part of the other; therefore the remaining angle BEC is double of the remaining angle BDC. Therefore the angle at the centre, &c. Q. E. D.

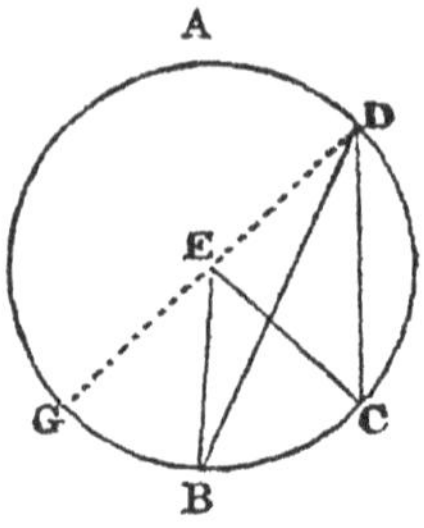

PROP. XXI. THEOR.

The angles in the same segment of a circle are equal to one another.

Let ABCD be a circle, and BAD, BED angles in the same segment BAED: The angles BAD, BED are equal to one another.

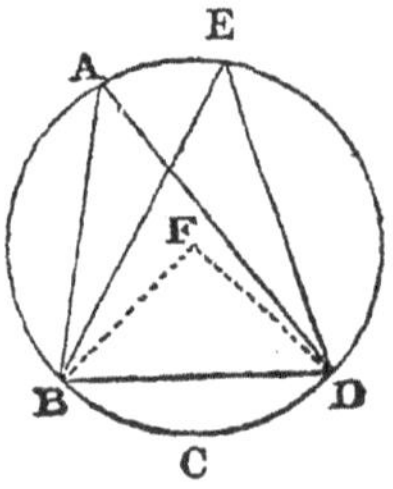

Take F the centre of the circle ABCD: And, first, let the segment BAED be greater than a semicircle, and join BF, FD: And because the angle BFD is at the centre, and the angle BAD at the circumference, and that they have the same part of the circumference, viz. BCD, for their base; therefore the angle BFD is double (III. 20.) of the angle BAD: For the same reason, the angle BFD is double of the angle BED: Therefore the angle BAD is equal to the angle BED.

But if the segment BAED be not greater than a semicircle, let BAD, BED be angles in it; these also are equal to one another: Draw AF to the centre, and produce it to C, and join CE: Therefore the segment BADC is greater than a semicircle; and the angles in it BAC, BEC are equal, by the first case: For the same reason, because CBED is greater than a semicircle, the angles CAD, CED are equal: Therefore the whole angle BAD is equal to the whole angle BED. Wherefore the angles in the same segment, &c. Q. E. D.

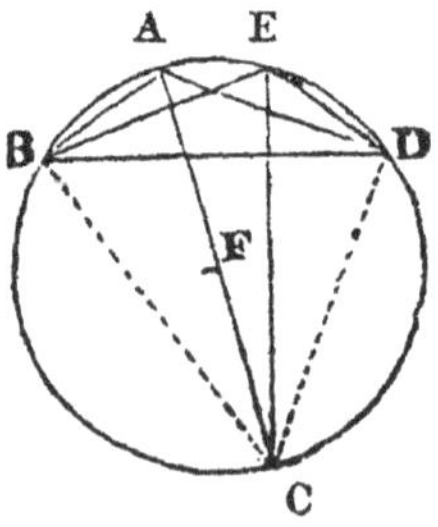

PROP. XXII. THEOR.

The opposite angles of any quadrilateral figure inscribed in a circle, are together equal to two right angles.

Let ABCD be a quadrilateral figure in the circle ABCD; any two of its opposite angles are together equal to two right angles.

Join AC, BD; and because the three angles of every triangle are equal (I. 32.) to two right angles, the three angles of the triangle CAB, viz. the angles CAB, ABC, BCA, are equal to two right angles: But the angle CAB is equal (III. 21.) to the angle CDB, because they are in the same segment BADC, and the angle ACB is equal to the angle ADB, because they are in the same segment ADCB: Therefore the whole angle ADC is equal to the angles CAB, ACB: To each of these equals add the angle ABC; therefore the angles ABC, CAB, BCA are equal to the angles ABC, ADC: But ABC, CAB, BCA are equal to two right angles; therefore also the angles ABC, ADC are equal to two right angles: In the same manner, the angles BAD, DCB may be shown to be equal to two right angles. Therefore, the opposite angles, &c. Q. E. D.

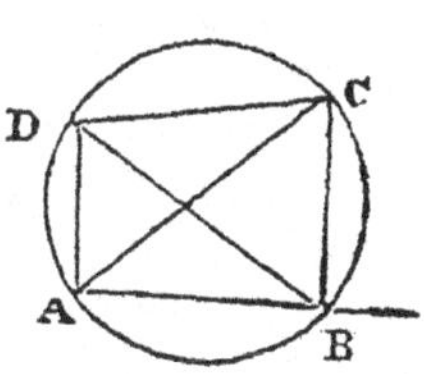

PROP. XXIII. THEOR.

Upon the same straight line, and upon the same side of it, there cannot be two similar segments of circles, not coinciding with one another.

If it be possible, let the two similar segments of circles, viz. ACB, ADB, be upon the same side of the same straight line AB, not coinciding with one another: Then, because the circle ACB cuts the circle ADB in the two points A, B, they cannot cut one another in any other point (III. 10.): One of the segments must therefore fall within the other: let ABC fall within ADB, and draw the straight line BCD, and join CA, DA: And because the segment ACB is similar to the segment ADB, and that similar segments of circles contain (III. Def. 11.) equal angles; the angle ACB is equal to the angle ADB, the

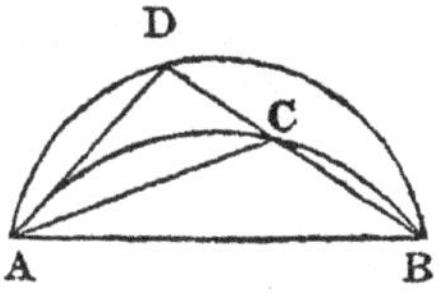

exterior to the interior, which is impossible (I. 16.). Therefore, there cannot be two similar segments of a circle upon the same side of the same line, which do not coincide. Q. E. D.

PROP. XXIV. THEOR.

Similar segments of circles upon equal straight lines are equal to one another.

Let AEB, CFD be similar segments of circles upon the equal straight lines AB, CD; the segment AEB is equal to the segment CFD.

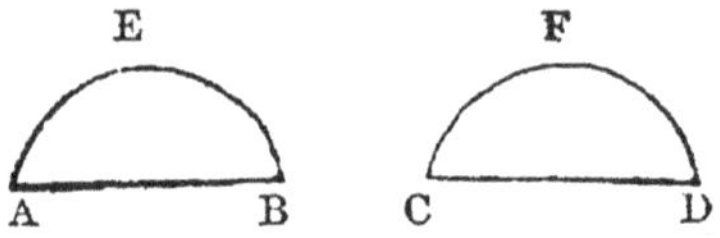

For if the segment AEB be applied to the segment CFD, so as the point A be on C, and the straight line AB upon CD, the point B shall coincide with the point D, because AB is equal to CD: Therefore the straight line AB coinciding with CD, the segment AEB must (III. 23.) coincide with the segment CFD, and therefore is equal to it. Wherefore similar segments, &c. Q. E. D.

PROP. XXV. PROB.

A segment of a circle being given, to describe the circle of which it is the segment.

Let ABC be the given segment of a circle; it is required to describe the circle of which it is the segment.

Bisect (I. 10.) AC in D, and from the point D draw (I. 11.) DB, at right angles to AC, and join AB: First, let the angles ABD, BAD be equal to one another; then the straight line BD is equal (I. 6.) to DA, and therefore to DC; and because the three straight lines DA, DB, DC, are all equal; D is the centre of the circle (III. 9.): From the centre D, at the distance of any of the three DA, DB, DC, describe a circle; this shall pass through the other points; and the circle of which ABC is a segment is described: And because the centre D is in

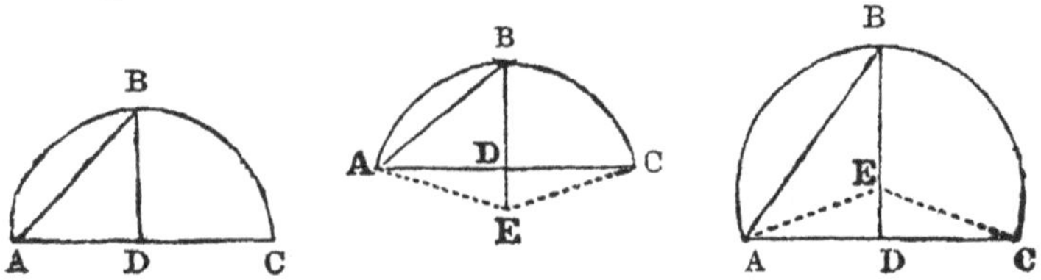

AC, the segment ABC is a semicircle; but if the angles ABD, BAD are not equal to one another, at the point A, in the straight line BA, make (I. 23.) the angle BAE equal to the angle ABD, and produce BD, if necessary, to E, and join EC; And because the angle ABE is equal to the angle BAE, the straight line BE is equal (I. 6.) to EA: And because AD is equal to DC, and DE common to the triangles ADE, CDE, the two sides AD, DE, are equal to the two CD, DE, each to each; and the angle ADE is equal to the angle CDE, for each of them is a right angle; therefore the base AE is equal (I. 4.) to the base EC: But AE was shown to be equal to BE, wherefore also BE is equal to EC: And the three straight lines AE, EB, EC are therefore equal to one another; wherefore (III. 9.) E is the centre of the circle. From the centre E at the distance of any of the three AE, EB, EC, describe a circle, this shall pass through the other points; and the circle of which ABC is a segment is described: And it is evident, that if the angle ABD be greater than the angle BAD, the centre E falls without the segment ABC, which therefore is less than a semicircle: But if the angle ABD be less than BAD, the centre E falls within the segment ABC, which is therefore greater than a semicircle: Wherefore a segment of a circle being given, the circle is described of which it is a segment. Which was to be done.

PROP. XXVI. THEOR.

In equal circles, equal angles stand upon equal circumferences, whether they be at the centres or circumferences.

Let ABC, DEF be equal circles, and the equal angles BGC EHF at their centres, and BAC, EDF at their circumferences: The circumference BKC is equal to the circumference ELF.

Join BC, EF; and because the circles ABC, DEF are equal, the straight lines drawn from their centres are equal: Therefore the two sides BG, GC, are equal to the two EH, HF;

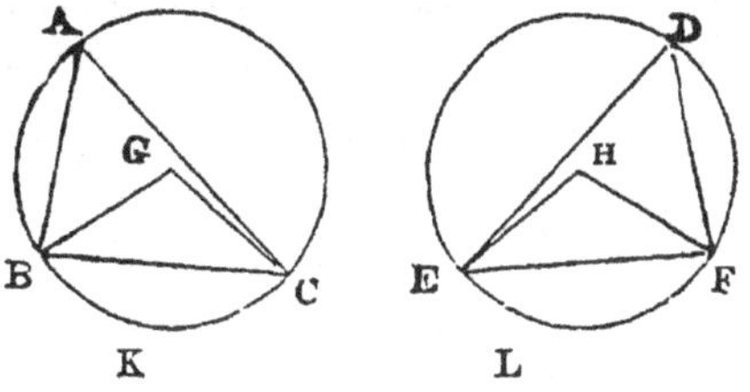

and the angle at G is equal to the angle at H; therefore the base BC is equal (I. 4.) to the base EF: And because the angle at A is equal to the angle at D, the segment BAC is similar (III. Def. 11.) to the segment EDF; and they are upon equal straight lines BC, EF; but similar segments of circles upon equal straight lines are equal (III. 24.) to one another, therefore the segment BAC is equal to the segment EDF: But the whole circle ABC is equal to the whole DEF; therefore the remaining segment BKC is equal to the remaining segment ELF, and the circumference BKC to the circumference ELF. Wherefore, in equal circles, &c. Q.E.D.

PROP. XXVII. THEOR.

In equal circles, the angles which stand upon equal circumferences are equal to one another, whether they be at the centres or circumferences.

Let the angles BGC, EHF at the centres, and BAC, EDF at the circumferences of the equal circles, ABC, DEF stand upon the equal circumferences BC, EF: The angle BGC is equal to the angle EHF, and the angle BAC to the angle EDF.

If the angle BGC be equal to the angle EHF, it is manifest (III. 20.) that the angle BAC is also equal to EDF. But if

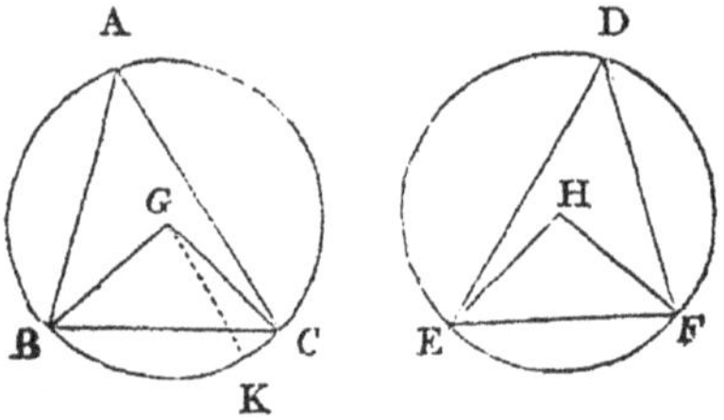

not, one of them is the greater: Let BGC be the greater, and at the point G, in the straight line BG, make the angle BGK equal to the angle EHF; but equal angles stand upon equal circumferences (III. 26.) when they are at the centre; therefore the circumference BK is equal to the circumference EF: But EF is equal to BC; therefore also BK is equal to BC, the less to the greater, which is impossible: Therefore the angle BGC is not unequal to the angle EHF; that is, it is equal to it: And the angle at A is half of the angle BGC, and the angle at D half of the angle EHF: Therefore the angle at

A is equal to the angle at D. Wherefore in equal circles, &c. Q.E.D.

PROP. XXVIII. THEOR.

In equal circles, equal straight lines cut off equal circumferences, the greater equal to the greater, and the less to the less.

Let ABC, DEF be equal circles, and BC, EF equal straight lines in them, which cut off the two greater circumferences BAC, EDF, and the two less BGC, EHF: the greater BAC is equal to the greater EDF, and the less BGC to the less EHF.

Take (III. 1.) K, L the centres of the circles, and join BK, KC, EL, LF: And because the circles are equal, the straight

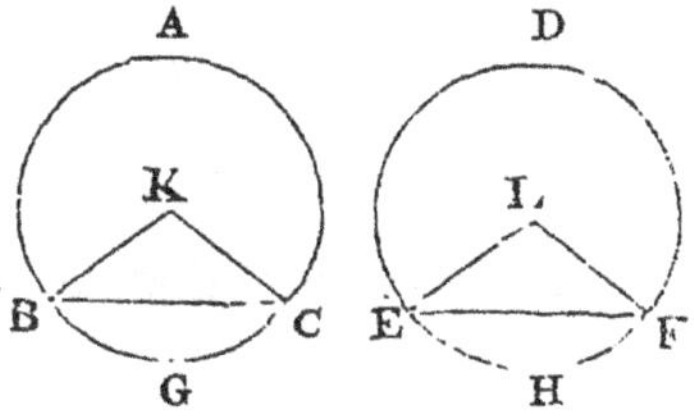

lines from their centres are equal; therefore BK, KC are equal to EL, LF; and the base BC is equal to the base EF; therefore the angle BKC is equal (I. 8.) to the angle ELF: But equal angles stand upon equal (III. 26.) circumferences, when they are at the centres; therefore the circumference BGC is equal to the circumference EHF. But the whole circle ABC is equal to the whole EDF; the remaining part therefore of the circumference, viz. BAC, is equal to the remaining part EDF. Therefore, in equal circles, &c. Q.E.D.

PROP. XXIX. THEOR.

In equal circles equal circumferences are subtended by equal straight lines.

Let ABC, DEF be equal circles, and let the circumferences BGC, EHF also be equal; and join BC, EF: The straight line BC is equal to the straight line EF.

Take (III. 1.) K, L the centres of the circles, and join BK, KC, EL, LF: and because the circumference BGC is equal to

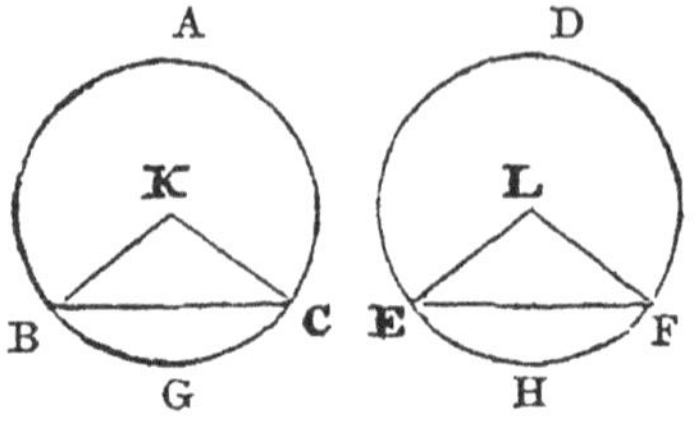

the circumference EHF, the angle BKC is equal (III. 27.) to the angle ELF: And because the circles ABC, DEF are equal, the straight lines from their centres are equal. Therefore BK, KC are equal to EL, LF, and they contain equal angles: Therefore the base BC is equal (I. 4.) to the base EF: Therefore, in equal circles, &c. Q.E.D.

PROP. XXX. PROB.

To bisect a given circumference, that is, to divide it into two equal parts.

Let ADB be the given circumference; it is required to bisect it.

Join AB, and bisect (I. 10.) it in C; from the point C draw CD at right angles to AB, and join AD, DB: The circumference ADB is bisected in the point D.

Because AC is equal to CB, and CD common to the triangles ACD, BCD, the two sides AC, CD are equal to the two BC, CD; and the angle ACD is equal to the angle BCD, because each of them is a right angle; therefore the base AD is equal (I. 4.) to the base BD. But equal straight lines cut off equal (III. 28.) circumferences, the greater equal to the greater, and the less to the less, and AD, DB are each of them less than a semicircle; because DC passes through the centre (III. 1. Cor.). Wherefore the circumference AD is equal to the circumference DB: Therefore the given circumference is bisected in D. Which was to be done.

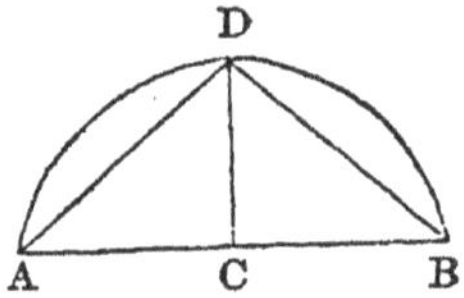

Prop. XXXI. Theor.

In a circle, the angle in a semicircle is a right angle; but the angle in a segment greater than a semicircle is less than a right angle; and the angle in a segment less than a semicircle is greater than a right angle.

Let ABCD be a circle, of which the diameter is BC, and centre E; and draw CA dividing the circle into the segments ABC, ADC, and join BA, AD, DC; the angle in the semicircle BAC is a right angle; and the angle in the segment ABC, which is greater than a semicircle, is less than a right angle; and the angle in the segment ADC, which is less than a semicircle, is greater than a right angle.

Join AE, and produce BA to F; and because BE is equal to EA, the angle EAB is equal (I. 5.) EBA; also, because AE is equal to EC the angle EAC is equal to ECA; wherefore the whole angle BAC is equal to the two angles ABC, ACB; But FAC, the exterior angle of the triangle ABC, is equal (I. 32.) to the two angles ABC, ACB; therefore the angle BAC is equal to the angle FAC, and each of them is therefore a right (I. Def. 10.) angle: Wherefore the angle BAC in a semicircle is a right angle.

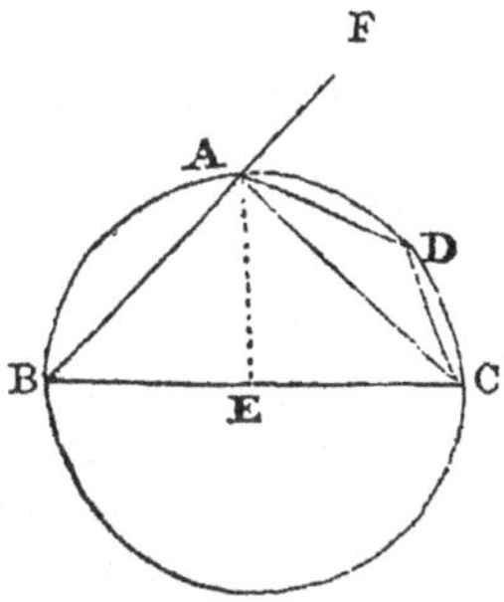

And because the two angles ABC, BAC of the triangle ABC are together less (I. 17.) than two right angles, and that BAC is a right angle, ABC must be less than a right angle; and therefore the angle in a segment ABC greater than a semicircle, is less than a right angle.

And because ABCD is a quadrilateral figure in a circle, any two of its opposite angles are equal (III. 22.) to two right angles; therefore the angles ABC, ADC are equal to two right angles; and ABC is less than a right angle; wherefore the other ADC is greater than a right angle.

Besides, it is manifest, that the circumference of the greater segment ABC falls without the right angle CAB, but the circumference of the less segment ADC falls within the right angle CAF. "And this is all that is meant, when in the Greek text, and the translations from it, the angle of the greater segment is said to be greater, and the angle of the less segment is said to be less, than a right angle.'

COR. From this it is manifest, that if one angle of a triangle be equal to the other two, it is a right angle, because the angle adjacent to it is equal to the same two; and when the adjacent angles are equal, they are right angles.

PROP. XXXII. THEOR.

If a straight line touches a circle, and from the point of contact a straight line be drawn cutting the circle, the angles made by this line with the line touching the circle, shall be equal to the angles which are in the alternate segments of the circle.

Let the straight line EF touch the circle ABCD in B, and from the point B let the straight line BD be drawn cutting the circle: The angles which BD makes with the touching line EF shall be equal to the angles in the alternate segments of the circle: that is, the angle FBD is equal to the angle which is in the segment DAB, and the angle DBE to the angle in the segment BCD.

From the point B draw (I. 11.) BA at right angles to EF, and take any point C in the circumference BD, and join AD, DC, CB; and because the straight line EF touches the circle ABCD in the point B, and BA is drawn at right angles to the touching line from the point of contact B, the centre of the circle is (III. 19.) in BA; therefore the angle ADB in a semicircle is a right (III. 31.) angle, and consequently the other two angles BAD, ABD are equal (I. 32.) to a right angle: But ABF is likewise a right angle; therefore the angle ABF is equal to the angles BAD, ABD: Take from these equals the common angle ABD; therefore the remaining angle DBF is equal to the angle BAD, which is in the alternate segment of the circle; and because ABCD is a quadrilateral figure in a circle, the opposite angles BAD, BCD are equal (III. 22.) to two right angles; therefore the angles DBF, DBE, being likewise equal (I. 13.) to two right angles, are equal to the angles BAD, BCD; and DBF has been proved equal to BAD: Therefore the remaining angle DBE

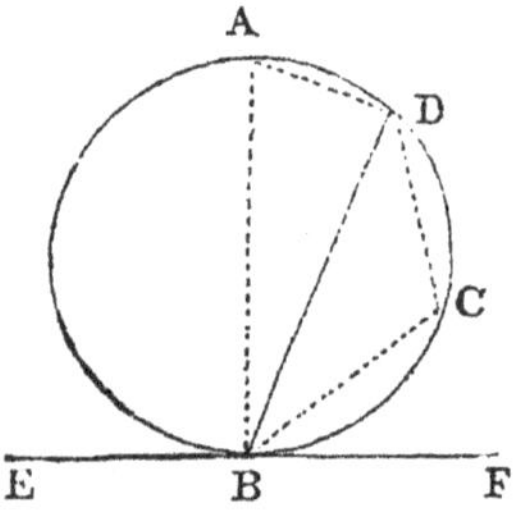

is equal to the angle BCD in the alternate segment of the circle.

Wherefore, if a straight line, &c. Q. E. D.

Prop. XXXIII. Prob.

Upon a given straight line to describe a segment of a circle, containing an angle equal to a given rectilineal angle.

Let AB be the given straight line, and the angle at C the given rectilineal angle; it is required to describe upon the given straight line AB a segment of a circle, containing an angle equal to the angle C.

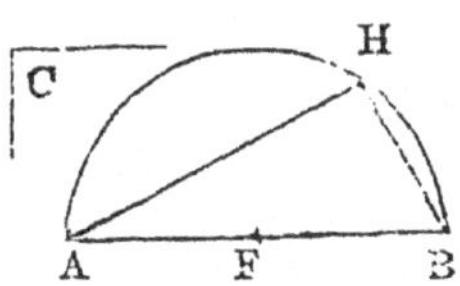

First, let the angle at C be a right angle, and bisect (I. 10.) AB in F, and from the centre F, at the distance FB, describe the semicircle AHB; therefore the angle AHB in a semicircle is (III. 31.) equal to the right angle at C.

But, if the angle C be not a right angle, at the point A, in the straight line AB, make (I. 23.) the angle BAD equal to the angle C, and from the point A draw (I. 11.) AE at right angles to AD; bisect (I. 10.) AB in F, and from F draw (I. 11.) FG at right angles to AB, and join GB: And because AF is equal to FB, and FG common to the triangles AFG, BFG, the two sides AF, FG are equal to the two BF, FG; and the angle AFG is equal to the angle BFG; therefore the base AG is equal (I. 4.) to the base GB; and the circle described from the centre G, at the distance GA, shall pass through the point

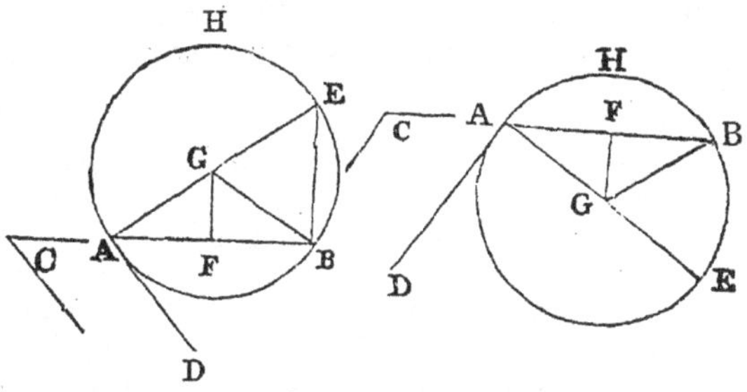

B; let this be the circle AHB: And because from the point A the extremity of the diameter AE, AD is drawn at right angles to AE, therefore AD (III. 16. Cor.) touches the circle;

and because AB drawn from the point of contact A cuts the circle, the angle DAB is equal to the angle in the alternate segment AHB (III. 32.). But the angle DAB is equal to the angle C, therefore also the angle C is equal to the angle in the segment AHB: Wherefore, upon the given straight line AB the segment AHB of a circle is described which contains an angle equal to the given angle at C. Which was to be done.

PROP. XXXIV. PROB.

To cut off a segment from a given circle which shall contain an angle equal to a given rectilineal angle.

Let ABC be the given circle, and D the given rectilineal angle; it is required to cut off a segment from the circle ABC that shall contain an angle equal to the angle D.

Draw (III. 17.) the straight line EF touching the circle ABC in the point B, and at the point B in the straight line BF make (I. 23.) the angle FBC equal to the angle D. Then, because the straight line EF touches the circle ABC, and BC is drawn from the point of contact B, the angle FBC is equal (III. 32.) to the angle in the alternate segment BAC of the circle:

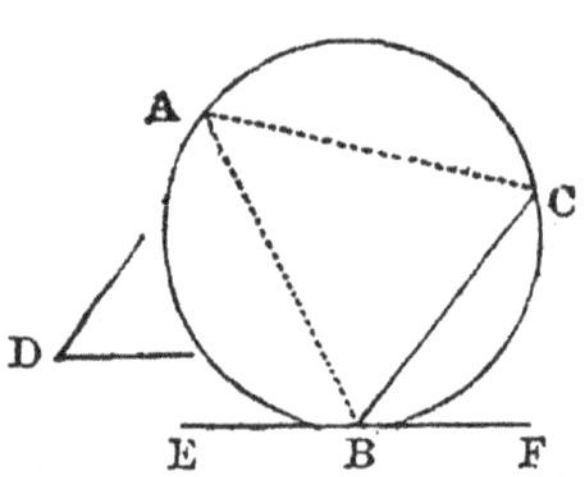

But the angle FBC is equal to the angle D; therefore the angle in the segment BAC is equal to the angle D: Wherefore the segment BAC is cut off from the given circle ABC containing an angle equal to the given angle D: Which was to be done.

PROP. XXXV. THEOR.

If two straight lines within a circle cut one another, the rectangle contained by the segments of one of them is equal to the rectangle contained by the segments of the other.

Let the two straight lines AC, BD, within the circle ABCD. cut one another in the point E· the rectangle contained by

AE, EC is equal to the rectangle contained by BE, ED.

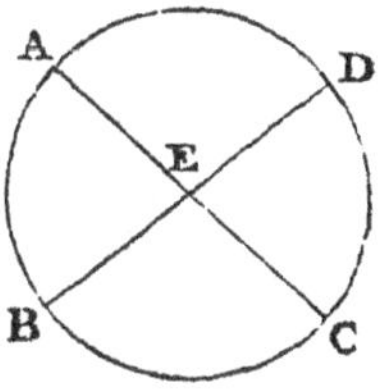

If AC, BD pass each of them through the centre, so that E is the centre; it is evident, that AE, EC, BE, ED, being all equal, the rectangle AE, EC is likewise equal to the rectangle BE, ED.

But let one of them BD pass through the centre, and cut the other AC, which does not pass through the centre, at right angles, in the point E: Then, if BD be bisected in F, F is the centre of the circle ABCD; join AF: And because BD, which passes through the centre, cuts the straight line AC, which does not pass through the centre, at right angles in E, AE, EC are equal (III. 3.) to one another: And because the straight line BD is cut into two equal parts in the point F, and into two unequal parts in the point E, the rectangle BE, ED together with the square of EF, is equal (II. 5.) to the square of FB; that is, to the square of FA; but the squares of AE, EF are equal (I. 47.) to the square of FA; therefore the rectangle BE, ED, together with the square of EF, is equal to the squares of AE, EF: Take away the common square of EF and the remaining rectangle BE, ED is equal to the remaining square of AE; that is, to the rectangle AE, EC.

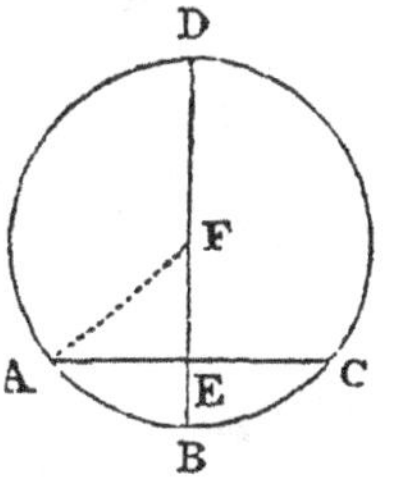

Next, let BD, which passes through the centre, cut the other AC, which does not pass through the centre, in E, but not at right angles: Then, as before, if BD be bisected in F, F is the centre of the circle. Join AF, and from F draw (I. 12.) FG perpendicular to AC; therefore AG is equal (III. 3.) to GC; wherefore the rectangle AE, EC, together with the square of EG, is equal (II. 5.) to the square of AG: To each of these squares add the square of GF; therefore the rectangle AE, EC, together with the squares of EG, GF, is equal to the squares of AG, GF: But the squares of EG, GF are equal to the square of EF; and the squares of AG, GF are equal to the square of AF: Therefore the rectangle AE, EC, together with the square of EF, is equal to the square of AF; that is, to the square of FB: But the square of FB is equal (II. 5.) to the

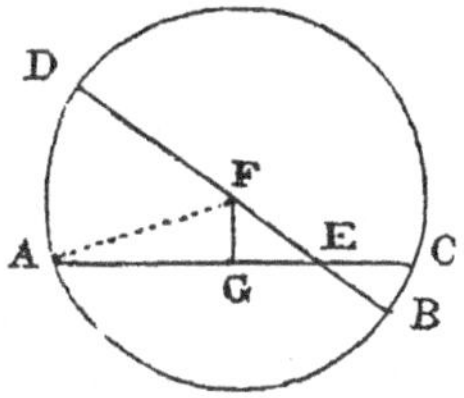

rectangle BE, ED, together with the square of EF; therefore the rectangle AE, EC, together with the square of EF, is equal to the rectangle BE, ED together with the square of EF: Take away the common square of EF, and the remaining rectangle AE, EC is therefore equal to the remaining rectangle BE, ED.

Lastly, Let neither of the straight lines AC, BD pass through the centre: Take the centre F, and through E, the intersection of the straight lines AC, DB, draw the diameter GEFH: And because the rectangle AE, EC is equal, as has been shown, to the rectangle GE, EH; and for the same reason, the rectangle BE, ED is equal to the same rectangle GE, EH; therefore the rectangle AE, EC is equal to the rectangle BE, ED. Wherefore, if two straight lines, &c. Q. E. D.

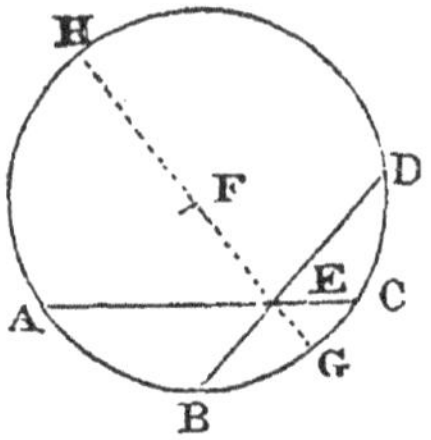

Prop. XXXVI. Theor.

If from any point without a circle two straight lines be drawn, one of which cuts the circle, and the other touches it; the rectangle contained by the whole line which cuts the circle, and the part of it without the circle, shall be equal to the square of the line which touches it.

Let D be any point without the circle ABC, and DCA, DB two straight lines drawn from it, of which DCA cuts the circle, and DB touches the same; The rectangle AD, DC is equal to the square of DB.

Either DCA passes through the centre, or it does not; first, let it pass through the centre E, and join EB; therefore the angle EBD is a right angle (III. 18.): And because the straight line AC is bisected in E, and produced to the point D, the rectangle AD, DC, together with the square of EC, is equal (II. 6.) to the square of ED, and CE is equal to EB: Therefore the rectangle AD, DC, together with the square of EB, is equal to the square of ED: But the square of ED is equal (I. 47.) to the squares of EB, BD, because EBD is a right angle: Therefore the rectangle AD, DC, together with the square of EB, is equal to the squares of EB, BD: Take

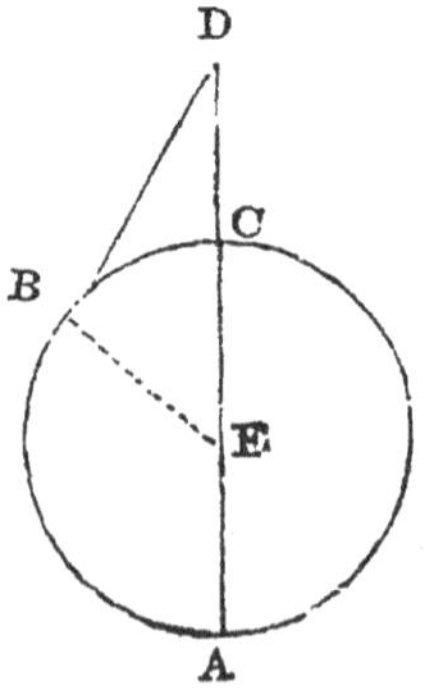

away the common square of EB; therefore the remaining rectangle AD, DC is equal to the square of the tangent DB.

But if DCA does not pass through the centre of the circle ABC, take (III. 1.) the centre E, and draw EF perpendicular (I. 12.) to AC, and join EB, EC, ED: And because the straight line EF, which passes through the centre, cuts the straight line AC, which does not pass through the centre, at right angles, it shall likewise bisect it (III. 3.); therefore AF is equal to FC: And because the straight line AC is bisected in F, and produced to D, the rectangle AD, DC, together with the square of FC, is equal (II. 6.) to the square of FD: To each of these equals add the square of FE; therefore the rectangle AD, DC, together with the squares of CE, FE, is equal to the squares of DF, FE: But the square of ED is equal (I. 47.) to the squares of DF, FE, because EFD is a right angle; and the square of EC is equal to the squares of CF, FE; therefore the rectangle AD, DC, together with the square of EC, is equal to the square of ED: And CE is equal to EB; therefore the rectangle AD, DC, together with the square of EB, is equal to the square of ED: But the squares of EB, BD are equal to the square (I. 47.) of ED, because EBD is a right angle; therefore the rectangle AD, DC, together with the square of EB, is equal to the squares of EB, BD: Take away the common square of EB; therefore the remaining rectangle AD, DC is equal to the square of DB. Wherefore, if from any point, &c. Q. E. D.

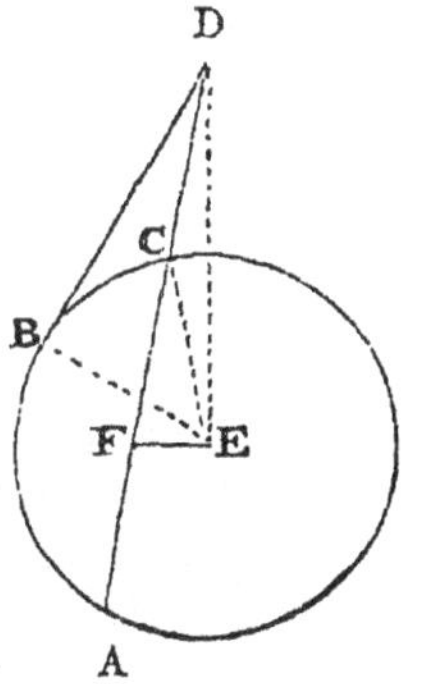

COR. If from any point without a circle, there be drawn two straight lines cutting it, as AB, AC, the rectangles contained by the whole lines and the parts of them without the circle are equal to one another, viz. the rectangle BA, AE to the rectangle CA, AF: for each of them is equal to the square of the straight line AD which touches the circle.

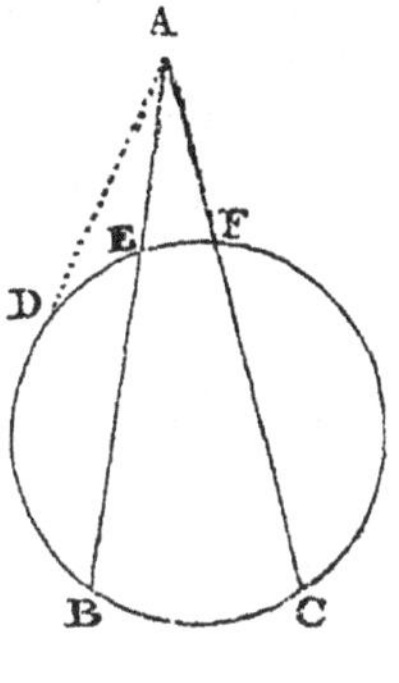

Prop. XXXVII. Theor.

If from a point without a circle there be drawn two straight lines, one of which cuts the circle, and the other meets it; if the rectangle contained by the whole line which cuts the circle, and the part of it without the circle, be equal to the square of the line which meets it, the line which meets shall touch the circle.

Let any point D be taken without the circle ABC, and from it let two straight lines DCA and DB be drawn, of which DCA cuts the circle, and DB meets it; if the rectangle AD, DC be equal to the square of DB; DB touches the circle.

Draw (III. 17.) the straight line DE touching the circle ABC, find its centre F, and join FE, FB, FD; and then FED is a right angle (III. 18.): And because DE touches the circle ABC and DCA cuts it, the rectangle AD, DC is equal (III. 36.) to the square of DE: But the rectangle AD, DC is, by hypothesis, equal to the square of DB: Therefore the square of DE is equal to the square of DB; and the straight line DE equal to the straight line DB: And FE is equal to FB, wherefore DE, EF are equal to DB, BF, each to each; and the base FD is common to the two triangles DEF, DBF; therefore the angle DEF is equal (I. 8.) to the angle DBF; but DEF is a right angle, therefore also DBF is a right angle: And FB, if produced, is a diameter, and the straight line which is drawn at right angles to a diameter, from the extremity of it, touches (III. 16.) the circle: Therefore DB touches the circle ABC. Wherefore, if from a point, &c. Q. E. D.

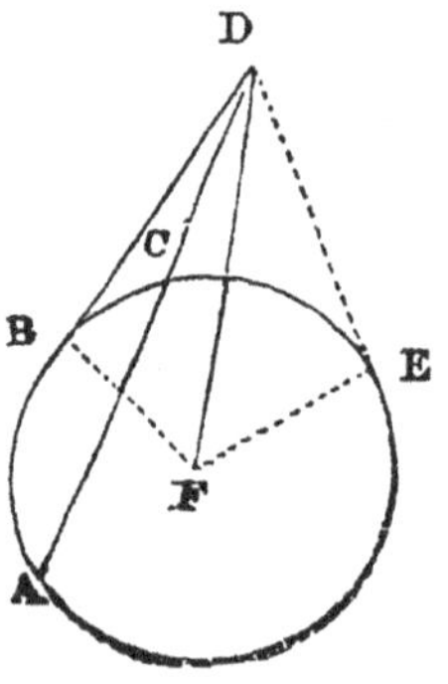

GEOMETRICAL EXERCISES.

EXERCISES ON BOOK I.

PROBLEMS. No. I.

1. *To trisect a right angle; that is, to divide it into three equal parts.*

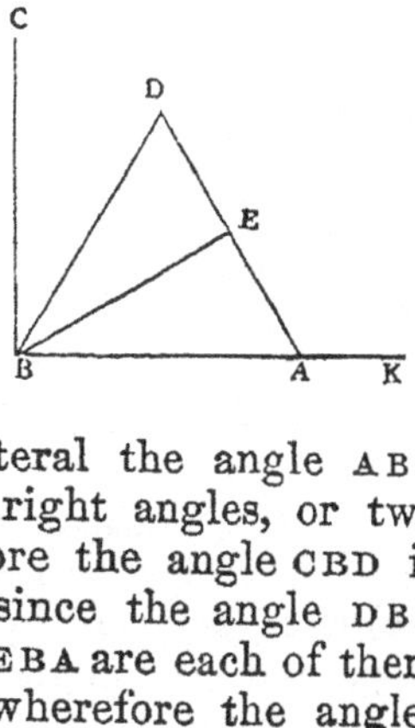

Let KBC be a right angle: It is required to trisect it.

In BK take any point A; on BA describe (Euclid I. 1.) the equilateral triangle ABD; and bisect (E. I. 9.) the angle ABD by the straight line BE; then the angles CBD, DBE, and EBA are equal to one another.

For the triangle ABD being equilateral the angle ABD (E. I. 5. and 32.) is one third of two right angles, or two thirds of the right angle ABC; therefore the angle CBD is one third of the right angle CBA; but since the angle DBA is bisected by BE, the angles DBE and EBA are each of them equal to one third of a right angle; wherefore the angles CBD, DBE, and EBA are equal to one another.

2. *To trisect a given straight line.*

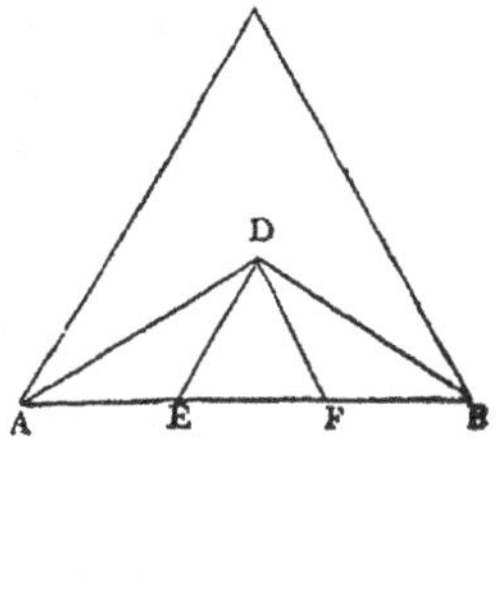

Let AB be the given straight line: It is required to divide it into three equal parts.

Upon AB describe (E. I. 1.) the equilateral triangle ABC; bisect (E. I. 9.) the angles CAB and CBA by the lines AD and BD meeting in D; and from D draw (E. I. 31.) DE parallel to CA, and DF parallel to CB: Then AB is trisected in the points E and F.

For since DE is parallel to CA and AD meets them, therefore (E. I. 29.) the angles DAC and ADE are equal; but (by constr.) the angle EAD is equal to the angle DAC; therefore the angle ADE is equal to the angle EAD; wherefore (E. I. 6.) AE is equal to ED. For the same reason FB is equal to FD. But DE being parallel to CA and DF to CB, the angle DEF is equal to the angle CAB, and DFE to CBA; and therefore the remaining angle EDF is equal to the remaining angle ACB; hence the triangle EDF is equiangular, and therefore ED, EF, FD are equal to one another; but ED has been shown to be equal to AE, and FD to FB, therefore AE, EF, FB are equal to one another.

3. *Through a given point to draw a straight line which shall make equal angles with two straight lines given in position.*

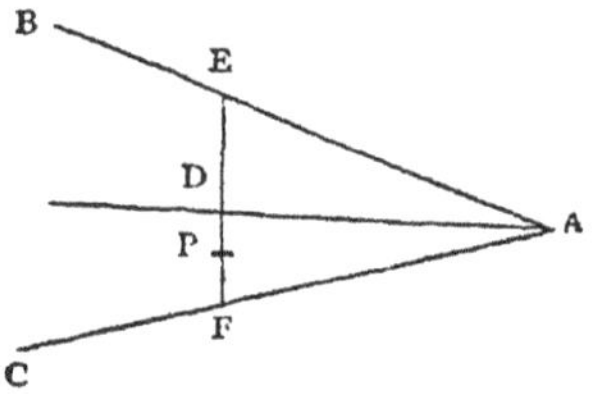

Let P be the given point, and BE and CF the lines given in position. It is required to draw, through the point P, a straight line which shall make equal angles with BE and CF.

Produce BE and CF to meet in A; bisect (E. I. 9.) the angle CAB by the line AD; and from P let fall (E. I. 12.) the perpendicular PD, and produce it on both sides to E and F. Then FE is the line required.

For, by construction, the angle DAE is equal to the angle DAF, and the angles at D are right angles, and moreover AD is common to both the triangles ADE and ADF, therefore (E. I. 26.) the angle AED is equal to the angle AFD.

4. *From two given points on the same side of a line given in position, to draw two lines which shall meet in that line, and make equal angles with it.*

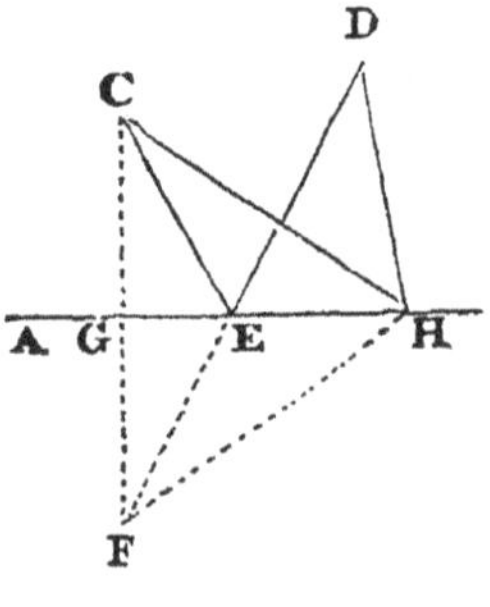

Let C and D be the two given points, and AH the line given in position.

From C let fall (E. I. 12.) the perpendicular CG, and produce it to F making GF equal to GC; join FD, CE; then CE and ED are the lines required.

Since CG is equal to GF, and GE is common to the two triangles CGE and FGE, and the angles at G are right angles, therefore (E. I. 4.) the triangles CGE and FGE are equal, and the angle CEG is equal to the angle FEG; but (E. I. 15.) the angle DEH is equal to the angle FEG, therefore the angle CEG is equal to the angle DEH.

5. On a given line to describe an isosceles triangle, whose perpendicular height is equal to the base.

6. To make an angle equal to half a right angle.

7. Given the diagonal of a square to construct it.

8. Given the base, the perpendicular height, and one of the angles at the base of a triangle, to construct it.

9. Trisect two right angles.

10. Divide a right angle into six equal parts.

11. Divide a right angle into four equal parts.

12. Given the hypotenuse of a right-angled triangle, and the difference of the two acute angles, to construct the triangle.

13. To construct a right-angled triangle, having given the hypotenuse, and one of the acute angles equal to one third a right angle. Show that the side opposite to the given angle is equal to one half the hypotenuse.

14. Given the two sides, and an angle opposite to one of the given sides of a triangle, to construct it. Show that, in general, there are two triangles answering the conditions of the problem.

15. Given the base, one of the other sides, and the perpendicular on the base of a triangle, to construct it.

16. On a given base, to construct an isosceles triangle having the angle at the vertex equal to one third a right angle.

17. From the vertex of a given triangle, to draw to the base a straight line which shall be less than the greater side by a given line.

18. Given the perpendicular and the equal side of an isosceles triangle, to construct it.

19. Through a given point P, to draw a straight line, which shall cut off equal parts from two straight lines AB and AC, cutting one another in A.

20 Given the vertical angle and the perpendicular height of an isosceles triangle, to construct it.

21. Given the base, the less angle at the base, and the difference of the sides of a triangle, to construct it.

22. From a given point, to draw the shortest line possible to a given straight line.

23. In a straight line given in position, but indefinite in length, to find a point, which shall be equidistant from each of two given points, on the same side of the given line, and in the same plane with it.

24. To draw a straight line from a given point to meet another straight line, which shall make with it an angle equal to a given rectilineal angle.

25. To find a point, in either of the equal sides of a given isosceles triangle, from which, if a straight line be drawn, perpendicular to that side, so as to meet the other side produced, it shall be equal to the base of the triangle.

26. From a given point to draw a straight line cutting two parallel straight lines, so that the part of it intercepted between them shall be equal to a given finite straight line, not less than the perpendicular distance of the two parallels.*

27. To trisect a triangle by lines drawn from the vertex to the base.

THEOREMS. No. 1.

1. *The straight line which joins the points of bisection of the two sides of a triangle is parallel to the base.*

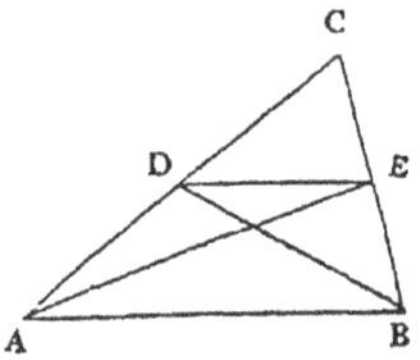

Let ABC be a triangle, D and E the points of bisection of the sides AC and BC respectively; then the straight line DE, which joins the points D and E, is parallel to the base AB.

Join the points A and E, B and D. Because AD is equal to DC, therefore (E. I. 38.) the triangle ABD is half the triangle ABC; and for the same reason the triangle ABE is also half the triangle ABC; therefore the triangles ABD and ABE are equal to one another; hence (E. I. 39.) DE is parallel to AB.

2. *If the sides of a trapezium be bisected, and the adjacent points be joined, the figure thus formed will be a parallelogram.*

* Should the student fail in solving any of these problems by the ordinary synthetic method, let him employ the method of analysis given at page 107.

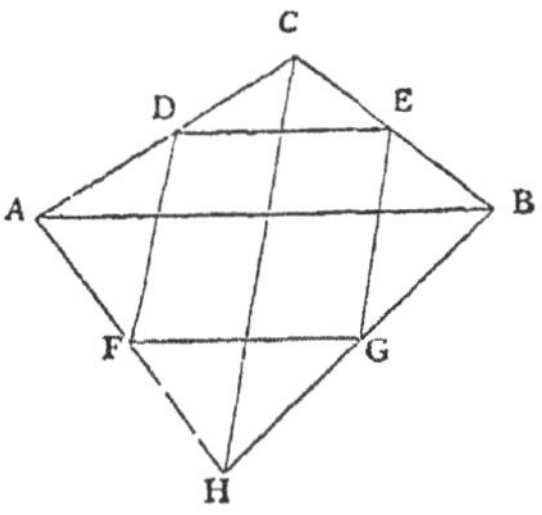

Let ACBH be a trapezium, whose sides are bisected in the points D, E, G and F; let the adjacent points be joined; then the figure DEGF is a parallelogram.

Draw the diagonals AB, CH. Since AC, BC are bisected in D and E (Theorem 1. No. 1.), therefore DE is parallel to AB; and for the same reason FG is parallel to AB; therefore (E. I. 30.) DE is parallel to FG. In the same way it may be shown that FD is parallel to GE; therefore the figure DEGF is a parallelogram.

3. Each of the angles of an equilateral triangle is equal to two-thirds of a right angle.

4. The diagonals of a parallelogram bisect one another.

5. The diagonals of a square cut each other at right angles.

6. The three sides of a triangle taken together exceed the double of any one side, and are less than the double of any two sides.

7. If two opposite sides of a quadrilateral figure be equal to one another, and the two remaining sides be also equal to one another, the figure is a parallelogram.

8. If the sides of a triangle are 3, 4, and 5, taken from a scale of equal parts, it is a right-angled triangle.

9. Any side of a triangle is greater than the difference between the other two sides.

10. Any one side of a quadrilateral figure is less than the sum of the remaining sides.

11. If either of the equal sides of an isosceles triangle be produced towards the vertex, the straight line which bisects the exterior angle is parallel to the base.

12. If the opposite angles of a quadrilateral figure be equal to each other, the figure is a parallelogram.

13. The diameters of an equilateral four-sided figure bisect one another at right angles.

14. The diameters of a rectangle are equal to one another.

15. The line which joins the points of bisection of any two sides of a triangle is equal to half the remaining side.

16. The diameters of any parallelogram divide it into four equal triangles.

17. The two triangles formed by drawing straight lines, from any point within a parallelogram, to the extremities of

either pair of opposite sides, are together half of the parallelogram.

18. The angle at the base of an isosceles triangle is equal to, or is less, or greater, than the half of the vertical angle, accordingly as the triangle is a right-angled, an obtuse-angled, or an acute-angled triangle.

19. If a perpendicular be drawn bisecting a given straight line, any point in this perpendicular is at equal distances from the extremities of the line.

20. Of all straight lines which can be drawn from a given point to an indefinite straight line, that which is nearer to the perpendicular is less than the more remote. And from the same point there cannot be drawn more than two straight lines equal to each other, viz. one on each side of the perpendicular.

21. The perimeter of an isosceles triangle is greater than the perimeter of a rectangular parallelogram, which is of the same altitude with and equal to the given triangle.

22. If from any point in the diameter (or diameter produced) of a parallelogram straight lines be drawn to the opposite angles, they will cut off equal triangles.

23. The two perpendiculars let fall from the extremities of the base of an isosceles triangle upon the sides, are equal to each other.

24. If a line be drawn bisecting a given angle, any point in that line will be equally distant from the two lines forming the angle.

25. Any straight line which bisects the diagonal of a parallelogram also bisects the parallelogram.

26. The perpendiculars drawn from the opposite angles of a parallelogram upon the diagonal are equal to one another.

27. If two straight lines bisect each other perpendicularly, the lines joining the extremities will form an equilateral parallelogram, or rhombus.

28. The area of a rhombus is equal to half the rectangle constructed upon the two diagonals of the rhombus.

29. Parallelograms whose sides and angles are equal are themselves equal.

30. The lines which join the middle points of the three sides of a triangle, divide it into four equal triangles.

31. The inscribed parallelogram DEGF, in Exercise 2., is equal to half the trapezium ACBH.

32. If perpendiculars be let fall from the angles at the

base of a triangle, upon the line drawn from the vertex to the middle of the base, these perpendiculars will be equal.

33. If two opposite sides of a trapezium be parallel to one another, the straight line joining their bisections, bisects the trapezium.

34. If from the extremities of the base of an isosceles triangle lines be drawn perpendicular to the sides, the angles made by these perpendiculars with the base are each equal to half the vertical angle.

35. In Theorem 1. No. 1., prove that the triangle DCE is one-half the triangle ABD.

PROBLEMS. No. 2.

1. *To draw a perpendicular to a straight line through its extremity without producing it.*

Let A be the extremity of the line AE: It is required to draw AC perpendicular to AE.

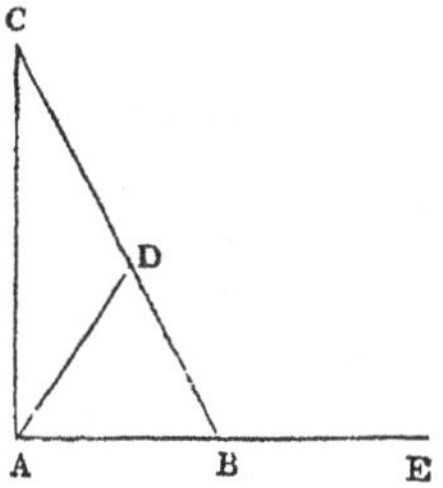

Take any portion AB, and on it construct the equilateral triangle ABD; produce BD to C, making DC equal to DB; join AC; then AC will be the perpendicular required.

Since AD is equal to DC (E. I. 5.) the angle DAC is equal to the angle DCA; and because ABD is an equilateral triangle the angle DAB is equal to the angle ABC; therefore the angles DAC and DAB taken together are equal to the sum of the angles DCA and ABC; but the angles DAC and DAB are together equal to the angle BAC; therefore the angle BAC is equal to the sum of the angles DCA and ABC; hence (E. I. 32.) the angle BAC is equal to a right angle.

2. *Given the sum of the three sides, and the angles at the base of a triangle, to construct it.*

Let AB be the sum of the three sides, and C and D the two given angles: It is required to describe a triangle, which shall have the sum of its sides equal to AB, and the angles at the base equal to C and D respectively.

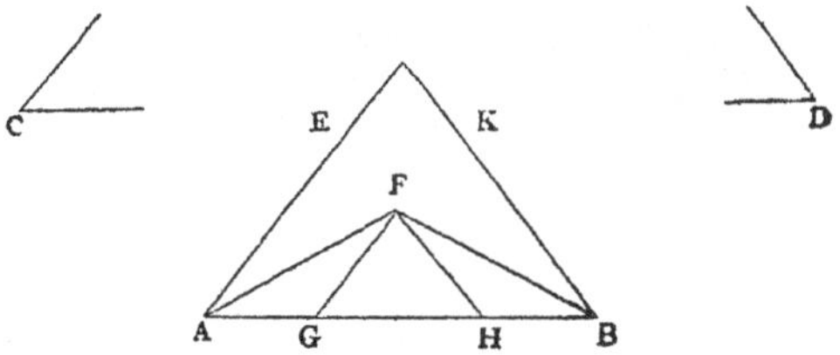

At the point A make (E. I. 23.) the angle BAE equal to the angle C, and at the point B the angle ABK equal to the angle D. Bisect (E. I. 9.) the angles BAE and ABK, by the lines AF and BF, meeting in F; through F draw (E. I. 31.) FG parallel to EA, and FH parallel to KB; then GHF is the triangle required.

For since FG is parallel to EA, therefore (E. I. 29.) the angle EAF is equal to the angle GFA; but (by construction) the angle EAF is equal to the angle GAF; therefore the angle GAF is equal to the angle GFA; therefore (E. I. 6.) GF is equal to GA. In the same manner it may be shown that HF is equal HB; hence the sum of the three sides of the triangle GHF is equal to the given straight line AB.

Again, since FG is parallel to EA, and FH to KB, therefore (E. I. 29.) the angle FGH is equal to the angle EAB, and the angle FHG to the angle KBA; but (by construction) the angle EAB is equal to the angle C, and the angle KBA is equal to the angle D; therefore also the angle FGH is equal to the angle C, and the angle FHG to the angle D. Wherefore, &c.

3. *To divide a given straight line into any given number of equal parts.*

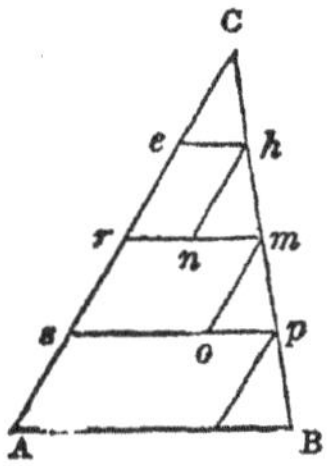

Let CB be a given straight line; it is required to divide it into any given number of equal parts.

From C draw an indefinite line CA making any convenient angle with CB; take any point *e* in CA, and make *er*, *rs*, *s*A, &c., each equal to C*e*, so that the number of these equal parts on CA, may be equal to the given number of parts into which CB is to be divided. Join the extremity A of the last part with the extremity B of the given line; and through the points *e*, *r*, *s*, &c. draw (E. I. 31.) *eh*, *rm*, *sp*, &c. parallel to AB, cutting CB in the points *h*, *m*, *p*, &c.; then these points will divide CB into the required number of equal parts.

Draw *hn*, *mo*, parallel to CA. Because *rm* is parallel to *eh* (E. I. 30.) and *hn* to *er*, therefore *ehnr* is a parallelogram, and (E. I. 34.) *hn* is equal to *er* or C*e*; the angle *nhm* is equal to the *e*C*h*, and the angle *hmn* to the angle C*he* (E. I. 29.); therefore (E. I. 26.) the triangle *nhm* is equal to the triangle *e*C*h*, and *hm* is equal to C*h*. In like manner it may be proved that *mp* is equal *hm*, and so on.

4. *To bisect a triangle by a line drawn from a given point in one of the sides.*

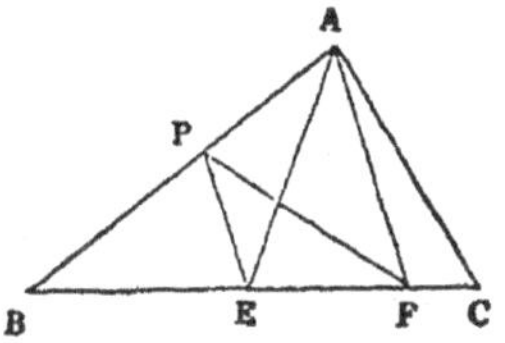

Let ABC be the triangle, and P the given point. Bisect BC in E; join AE, PE; from A draw AF parallel to PE; and join PF; then PF will bisect the triangle ABC.

Since AF is parallel to PE, therefore (E. I. 37.) the triangle PEA is equal to the triangle PEF. To these equals add the triangle BEP, therefore the triangle ABE is equal to the triangle BPF; but since BE is equal to EC, therefore (E. I. 38.) the triangle ABE is equal to half the triangle ABC; therefore the triangle BPF is also equal to half the triangle ABC.

5. Given the base, the sum of the remaining sides, and one of the angles at the base of a triangle, to construct it.

6. Given the base and the vertical angle of an isosceles triangle, to construct it.

7. To construct a triangle from three parts given, of which one, at least, is a side.

8. Upon a given base to describe an isosceles triangle equal to a given rectangle.

9. To divide a triangle into any given number of equal parts by lines drawn from the vertex to the base.

10. Describe a square which shall be equal to the difference between two given squares.

11. Describe a square which shall be equal to the sum of three given squares.

12. Construct a triangle which shall have the magnitude of its angles as the numbers 1, 2, and 3.

13. To find a point within a triangle, so that lines drawn to the angles shall divide the triangle into three equal parts.

14. Draw a line EF parallel to the base BC of a triangle ABC, so that EF shall be equal to BE.

15. Construct a parallelogram, having given the diagonals and the angle they make with each other.

16. Trisect a given angle, which is the half, or the quarter, or the eighth part, and so on, of a right angle.

17. In a given square to inscribe an equilateral triangle, having one of its angular points upon one of the angular points of the square, and its two remaining angular points one in each of two adjacent sides of the square.

18. To bisect a parallelogram by a line drawn through a given point in one of its sides.

19. To bisect a trapezium (1st) by a line drawn from one of its angular points, (2d) by a line drawn from a given point in one of its sides.

20. A plane rectilineal figure of any number of sides being given, to find an equal rectilineal figure which shall have the number of its sides less, by one, than that of the given figure.

21. To trisect a parallelogram by lines drawn from a given point in one of its sides.

22. To describe a parallelogram, the surface and perimeter of which shall be respectively equal to the surface and perimeter of a given triangle.

23. To describe a parallelogram which shall be of a given altitude, and equiangular with a given parallelogram, and also equal to it.

24. From the circumference of a given circle to draw to a straight line given in position, a line which shall be equal and parallel to a given straight line.

25. To describe a triangle which shall be equal to a given equilateral pentagon, and of the same altitude.

26. To inscribe a square in a given right-angled isosceles triangle.

27. To inscribe a square in a given equilateral four-sided figure.

28. To inscribe a square in a given quadrant of a circle.

29. Given one angle, a side opposite to it, and the difference of the other two sides, to construct the triangle.

30. Given the perimeter and the vertical angle of an isosceles triangle, to construct it.

31. From one of the angles of a parallelogram to draw a line to the opposite side, which shall be equal to that side together with the segment of it which is intercepted between the line and the opposite angle.

32. Two straight lines are given in position, without being produced to meet; it is required to draw a line which would (if produced) bisect the angle between the two straight lines.

33. Given the vertical angle, the difference of the angles at the base, and the perpendicular height of a triangle, to construct it.

34. In a given triangle to construct an equilateral paral-

lelogram, which shall have one of its angles coinciding with one of the angles of the triangle, and the opposite angular point situated in one of the sides of the triangle.

THEOREMS. No. II.

1. *The line joining the vertex and the middle of the base of a triangle, bisects every line that is drawn parallel to the base, and is terminated by the two remaining sides of the triangle.*

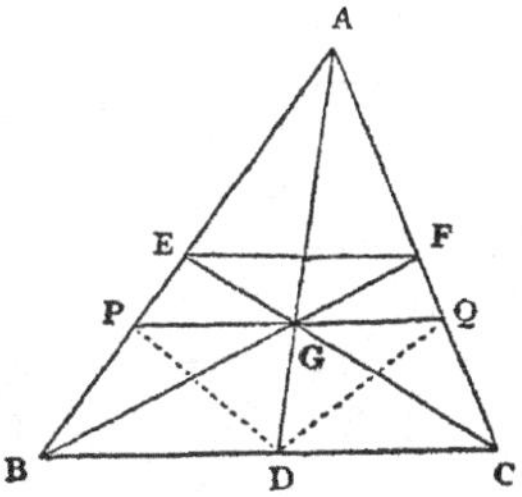

Let PQ be any straight line, drawn parallel to the base BC of the triangle ABC; and let AD, joining the vertex and the middle D of BC, cut PQ in G; then PQ is bisected in the point G.

If PG be not equal to GQ, one of them is the greater: let PG be the greater; and join DQ, and DP.

Since (hyp.) BD is equal to DC, and PQ is parallel to BC, therefore (E. I. 38.) the triangle BDP is equal to the triangle DCQ; also because BD is equal to DC, therefore the triangle BDA is equal to the triangle CDA; therefore, taking the former equal triangles from the latter, there remains the triangle DPA equal to the triangle DQA. But since PG is greater than GQ, the triangle APG is greater than AGQ, and the triangle GPD is greater than GQP; therefore, the whole triangle DPA is greater than the whole triangle DQA; but it has been shown that the triangle DPA is equal to the triangle DQA; and it is also greater, which is absurd: hence neither of the two lines PG, GQ, can be greater than the other; therefore PG must be equal to GQ.

2. *The two sides of a triangle are together greater than the double of the line joining the vertex and the middle of the base.*

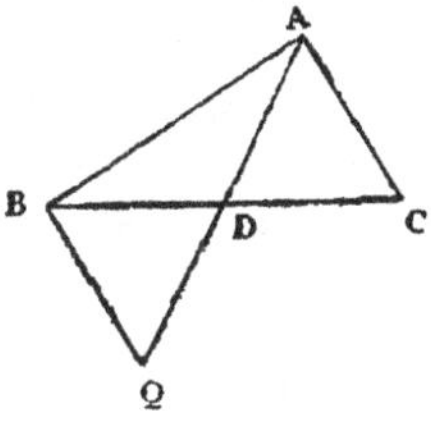

Let ABC be a triangle, and AD the straight line joining the vertex A, and the middle D, of the base BC; then AB and AC are together greater than twice AD. Produce AD to Q making DQ equal to AD; and join BQ.

Since (*hyp.*) BD is equal to DC, and (*constr.*) DQ is equal to AD, and (E. I. 15.)

the angle BDQ is equal to ADC; therefore (E. I. 4.) BQ is equal to AC. But (E. I. 20.) AB, BQ are together greater than AQ; but AC has been proved to be equal to BQ, and AQ (by const.) is the double of AD; therefore AB, AC are together greater than the double of AD.

3. If either diagonal of a parallelogram be equal to one of the sides, the other diagonal will be greater than any side of the parallelogram.

4. A trapezium, having two of its sides parallel, is equal to half of a rectangle between the same parallels, and having its base equal to the sum of the two parallel sides of the trapezium.

5. If the side of a square be equal to the diagonal of another square, the former square is double of the latter.

6. The two sides of a triangle are together greater than the double of the line drawn from the vertex to the base, bisecting the vertical angle.

7. If in the sides of a square, at equal distances from the four angles, four other points be taken, one in each side; the figure contained by the straight lines which join them shall also be a square.

8. If four straight lines cut each other, without including a space, but so as to make three internal angles, towards the same parts, together less than four right angles, the two lines which are not joined shall meet, if produced.

9. If two opposite sides of a parallelogram be bisected, and lines be drawn from these two points of bisection to the opposite angles, the lines will be parallel, and will trisect the diagonal.

10. In Prob. 4. No. 1. the sum of the lines CE and ED is less than the sum of any other two lines which can be drawn from the given points C and D to meet the line AH.

11. The perimeter of an isosceles triangle is less than that of any other equal triangle on the same base.

12. If from the angular points of the squares described upon the sides of a right-angled triangle perpendiculars be let fall upon the hypotenuse produced, they will cut off equal segments, and the perpendiculars will together be equal to the hypotenuse.

13. If a line PQ be drawn parallel to the base BC of a triangle ABC (see fig. Theo. 1. No. 2.) through the point G where the lines CG and BG bisecting the angles of the base

intersect, the line PQ will be equal to the sum of the lines BP and CQ.

14. The difference of the angles at the base of a triangle is double the angle contained by a line drawn from the vertex perpendicular to the base, and another line bisecting the angle at the vertex.

15. If in two opposite sides of a parallelogram two points be taken, one in each of those sides equidistant from two opposite angles of the figure, and if two other points be likewise taken, in the two other opposite sides, equidistant from the same two angles, the figure contained by the straight lines joining the four points so taken shall be a parallelogram.

16. The area of a trapezium is half that of a rectangle, whose base is one of the diagonals of the trapezium, and altitude the sum of the perpendiculars let fall upon the diagonal.

17. The lines which bisect the angles of a parallelogram, form a rectangle.

18. Two sides of a triangle being given, its area will be greatest when the given sides contain a right angle.

19. If the two exterior angles at the base of a triangle be bisected, and the bisecting lines produced until they intersect, the line drawn from this point to the vertical angle will bisect it.

20. In the figure E. I. 47. prove that, if BG and CH be joined, these lines will be parallel.

EXERCISES ON BOOK II.

1. THEOREM. *In any triangle, if a line be drawn from the vertex bisecting the base, the sum of the squares of the two sides of the triangle is double the sum of the squares of the bisecting line and of half the base.*

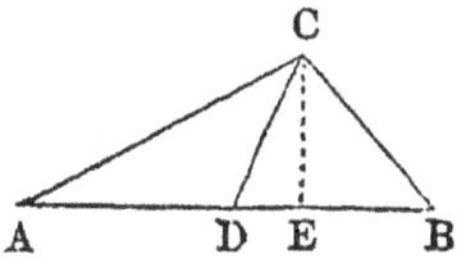

Let ABC be a triangle; and from the vertex C let CD be drawn to the middle, D, of the base; then the squares of AC, BC are together double the squares of DC, AD.

From C draw CE perpendicular to AB.

Then (E. II. 12.) the square of AC is greater than the squares of DC, AD, by twice the rectangle AD, DE; and (E. II. 13.) the square of BC is less than the squares of DC, BD by twice the rectangle BD, DE ; therefore the squares of AC, BC are equal to twice the square of DC, and twice the square of AD, since AD is equal to BD.

Or thus.

$$\text{(E. II. 12.)} \quad AC^2 = DC^2 + AD^2 + 2\,AD.DE,$$
$$\text{and (E. II. 13.)} \quad BC^2 = DC^2 + BD^2 - 2\,BD.DE$$
$$= DC^2 + AD^2 - 2\,AD.DE,$$
$$\therefore AC^2 + BC^2 = 2\,DC^2 + 2\,AD^2.$$

2. The square upon the whole line is equal to four times the square upon half the line.

3. The rectangle under any two lines is equal to three times the rectangle under either of them and a third of the other.

4. If two lines be divided, each into any number of parts, the rectangle contained by the two lines, is equal to the sum of the rectangles contained by the several parts of the one and the several parts of the other.

5. In any triangle if a line be drawn from the vertex at right angles to the base, the difference of the squares of the sides is equal to the difference of the squares of the segments of the base.

6. If from one of the acute angles of a right-angled triangle, a line be drawn to the opposite side, the squares of that side and the line so drawn are together equal to the squares of the segment adjacent to the right angle and of the hypotenuse.

7. In any isosceles triangle, if a line be drawn from the vertex to any point in the base, the square upon this line, together with the rectangle contained by the segments of the base, is equal to the square upon either of the equal sides.

8. The rectangle contained by the sum and the difference of two lines is equal to the difference of their squares.

9. If a line be divided into five equal parts, the square of the whole line is equal to the square of the line which is made up of four of those parts, together with the square of the line which is made up of three of those parts.

10. The squares of the sides of a parallelogram are together equal to the squares of its diameters taken together.

11. If from the three angles of a triangle lines be drawn to the points of bisection of the opposite sides, four times the sum of the squares of these lines is equal to three times the sum of the squares of the sides of the triangle.

12. If from any point within a rectangle lines be drawn to,

the angular points, the sums of the squares of those which are drawn to the opposite angles are equal.

13. The square of the base of an isosceles triangle is double the rectangle contained by either side and by the line intercepted between the perpendicular let fall upon it from the opposite angle, and the extremity of the base.

14. If two sides of a trapezium be parallel to each other, the squares of its diagonals are together equal to the squares of its two sides, which are not parallel, and twice the rectangle contained by its parallel sides.

15. The sum of the perpendiculars let fall from any point within an equilateral triangle, upon the three sides, is equal to the perpendicular let fall from one of the angles upon the opposite side.

16. The squares of the diagonals of a trapezium are together double the squares of the two lines joining the bisections of the opposite sides.

17. The squares of the diagonals of a trapezium are together less than the squares of the four sides, by four times the square of the line joining the points of bisection of the diagonals.

18. If from any point within (or without) a trapezium perpendiculars be let fall on every side, the sum of the squares of the alternate segments made by them will be equal.

19. To divide a given line into two parts, such that the squares described upon them shall be equal to a given square. Show when the problem becomes impossible.

20. Divide a given straight line into two parts, such that twice the rectangle contained by them may be equal to the square of one of the parts.

21. Divide a given straight line into two parts, such that nine times the square described upon one part may be equal to the square described upon the other.

22. Divide a given straight line into two parts, such that the square of the whole line may be equal to the square of one of the parts, together with four times the rectangle contained by the two parts.

23. Divide a given straight line into three parts, such that the square upon the sum of the greatest and least parts may be four times the square upon the remaining part; and the sum of the squares upon the two least parts may be equal to the square upon the greatest.

24. Upon a given line, as an hypotenuse, to describe a

right-angled triangle, such that the hypotenuse, together with the less of the two sides, shall be double the greater.

25. Show that the algebraical proposition

$$(a+b)^2 + (a-b)^2 = 2a^2 + 2b^2,$$

is equivalent to Props. 9 and 10 of the Second Book of Euclid.

26. Find algebraical propositions equivalent to Propositions 1, 2, 3, 4, 5, 6, 7, and 8 of the Second Book of Euclid.

EXERCISES ON BOOK III.

Problems.

1. *To describe a circle which shall touch a given straight line in a given point, and also touch a given circle.*

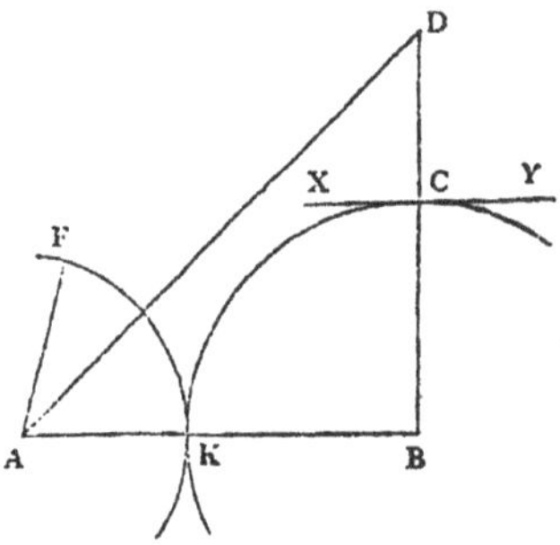

Let FK be the given circle, and C the given point in the straight line XY: It is required to describe a circle which shall touch XY in the point C, and also touch the circle FK.

Through C draw (E. I. 11.) the line DCB perpendicular to XY; find (E. III. 1.) the centre A of the circle FK, and draw any radius AF; make CD equal to AF, and join A, D; from A draw (E. I. 23.) AB, making the angle DAB equal to the angle ADB, and let DCB meet AB in the point B. Then B is the centre of the circle required.

Since the angle DAB is equal to the angle ADB (constr.), therefore (E. I. 6.) BD is equal to BA; and CD is equal to AF or AK; therefore the remainder BC is equal to the remainder BK, and consequently a circle described from B as a centre, with the radius BC, will pass through K, and (E. III. 16.) it will touch XY in C, and moreover it will touch the circle FK in K, for a line drawn perpendicular at K will form a common tangent to the circles FK and CK.

2. *Given the base, the vertical angle, and the perpendicular height of a triangle, to construct it.*

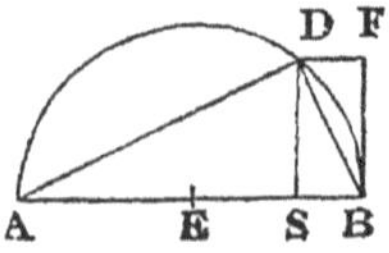

Let AB be the given base: Upon AB describe (E. III. 33.) a segment of a circle ADB, which shall contain an angle equal to the given vertical angle; from B draw BF perpendicular to AB, and make BF equal to the given height of the triangle; through F draw FD parallel to AB, meeting the arc of the circle in D; join B, D and A, D: Then ABD is the triangle required.

Draw DS parallel to FB, then BFDS is a parallelogram, and (E. I. 34.) DS is equal to FB; and (constr.) because the angle SBF is a right angle, therefore (E. I. 29.) the angle DSA is a right angle, that is, DS is perpendicular to AB, and it has been shown to be equal to FB, which was taken equal to the given perpendicular; moreover (constr.) the angle ADB is equal to the given vertical angle; therefore ADB is the triangle which was to be constructed.

3. *To determine the position of a point of observation,* D, *at which lines drawn from three objects* (A, B, C), *whose distances from each other are given, shall make with each other angles equal to given angles; that is, let the angles* ADB *and* ADC *be given angles.*

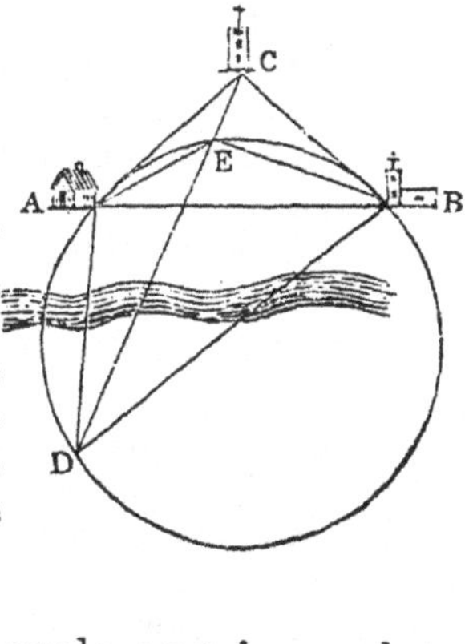

Let A, B, C be the three given points or objects; join AB, and on it describe (E. III. 33.) a segment of a circle ADB, which shall contain an angle equal to the angle which the lines drawn from A and B are to include. Describe the whole circle ADBE, and make (E. I. 23.) the angle ABE equal to the angle which the lines drawn from A and C are to include. Join C, E; and produce it to D, the circumference of the circle ADBE; then D is the position of the point required.

Join A, E; then (E. III. 21.) the angle ADC is equal to the angle ABE, which was made equal to the angle formed by the lines drawn from A and C; and, moreover, the angle ADB is, by construction, equal to the angle formed by the lines drawn from A and B.

4. Through a given point within a circle, to draw a chord which shall be bisected in that point.

5. To draw a tangent to a circle, which shall be parallel to a given straight line.

6. Given the hypotenuse and the perpendicular let fall upon it from the opposite angle, to construct the right-angled triangle.

7. Divide the circumference of a circle into six equal parts.

8. From two given points on the same side of a line given in position, to draw two straight lines which shall contain a given angle, and be terminated in that line.

9. To describe a circle which shall have a given radius and its centre in a given line, and shall also touch another line, inclined to the former at a given angle.

10. To describe a circle which shall pass through a given point, and touch a given line in another given point.

11. To describe a circle which shall pass through a given point, and touch a given circle in another given point.

12. To describe a circle which shall have a given radius, and which shall touch two given lines, not parallel to each other.

13. To describe a circle, which shall touch a straight line in a given point, and also touch a given straight line.

14. To describe two circles, each having a given radius, which shall touch each other, and the same given straight line on the same side of it.

15. To describe three circles of equal diameters, which shall touch each other.

16. To describe a circle which shall pass through a given point, have a given radius, and touch a given straight line.

17. To describe a circle which shall pass through two given points, and touch a given straight line.

18. To describe a circle the centre of which may be in the perpendicular of a given right-angled triangle, and the circumference pass through the right angle and touch the hypotenuse.

19. Through a given point to draw a straight line, the part of which intercepted by the circle, shall be equal to a given line, which is not greater than the diameter.

20. To draw a straight line which shall touch two given circles.

21. In the diameter of a circle produced, to determine a point from which a tangent drawn to the circumference shall be equal to the diameter.

22. Two circles being given in magnitude and position, to find a point in the circumference of one of them, to which if a tangent be drawn cutting the circumference of the other,

the part intercepted between the two circumferences may be equal to a given line.

23. Upon a given straight line to describe a segment of a circle, which shall be similar to a given segment of another circle.

24. To divide a given circular arch into two parts, so that the sum of their chords may be equal to a given straight line, greater than the chord of the whole arch, but not greater than the double of the chord of half the arch.

25. The perimeter, the vertical angle, and the altitude of a triangle being given, to construct it.

26. Two circles being given in position and magnitude, to draw a straight line cutting them so that the chords in each circle may be equal to a given line, not greater than the diameter of the smaller circle.

27. Having given the radii of two circles which cut each other, and the distance between their centres, to draw a line of given length through their point of intersection, so as to terminate in their circumferences.

28. To describe a rectangle which shall be equal to a given square, and have its adjacent sides together equal to a given line.

29. Having given the distance of the centres of two equal circles which cut each other, to inscribe a square in the space included between the two circumferences.

30. In a given circle to inscribe a rectangle equal to a given rectilineal figure.

31. To inscribe a circle in a given quadrant.

32. The vertical angle, the base, and the sum of the three sides of a triangle being given, to construct it.

33. The vertical angle, the base, and the excess of the greater of the two remaining sides, of a triangle, above the less, being given, to construct the triangle.

34. To produce a given line, so that the rectangle contained by the whole line thus produced, and the part of it produced, shall be equal to a given square.

35. From the obtuse angle of an obtuse-angled triangle, to draw a straight line to the base, the square of which shall be equal to the rectangle contained by the segments, into which it divides the base.

36. A flag-staff having a given height stands on the top of a tower whose height is also given; at what point in the

horizontal line, drawn from the foot of the tower, will the flag-staff appear under the greatest angle?

37. Given the vertical angle, the difference of the two sides containing it, and the difference of the segments of the base made by a perpendicular from the vertex; to construct the triangle.

Theorems.

1. If two circles cut each other, the straight line joining their two points of intersection is bisected, at right angles, by the straight line joining their centres.

2. If two isosceles triangles be of equal altitudes, and the side of one be equal to the side of the other, their bases shall be equal.

3. Any two chords of a circle, which cut a diameter in the same point and at equal angles, are equal to one another.

4. If a straight line be drawn to touch a circle and parallel to a chord, the point of contact will be the middle point of the arc cut off by that chord.

5. If from a point without a circle, two straight lines be drawn to the concave part of the circumference, making equal angles, with a line joining the same point and the centre, the parts of the lines which are intercepted within the circle are equal.

6. If a circle be described on the radius of another circle, any straight line drawn from the point where they meet to the outer circumference, is bisected by the interior one.

7. The two straight lines in a circle, which join the extremities of two parallel chords, are equal to each other.

8. The straight lines joining the extremities of the chords of two equal arches of the same circle, towards the same parts, are parallel to each other.

9. If two circles cut each other, and from either point of intersection diameters be drawn, the extremities of these diameters and the other point of intersection shall be in the same straight line.

10. If a straight line touch the interior of two concentric circles and be placed in the exterior one, it will be bisected in the point of contact.

11. If any number of equal straight lines be placed in a circle, the locus of their points of bisection will be a circle.

12. If from any two points in the circumference of the greater of two given concentric circles, two straight lines be drawn so as to touch the less circle, they shall be equal to one another.

13. If a quadrilateral rectilineal figure be described about a circle, the angles subtended, at the centre of the circle, by any two opposite sides of the figure, are, together, equal to two right angles.

14. If two circles touch each other externally or internally, two straight lines drawn through the point of contact, will intercept arcs the chords of which are parallel.

15. If two equal circles cut each other, and from either point of intersection a circle be described cutting them, the points where this circle cuts them, and the other point of intersection of the equal circles, are in the same straight line.

16. If a straight line touch the interior of two concentric circles, and be terminated both ways by the circumference of the outer circle, its squares shall be equal to the difference of the squares of the diameters of the circles.

17. The perpendiculars let fall from the three angles of any triangle upon the opposite sides, intersect each other in the same point.

18. If two opposite angles of a quadrilateral figure be together equal to two right angles, a circle may be described about it.

19. If from the extremities of the diameter of a circle tangents be drawn, and produced to intersect a tangent to any point of the circumference; the straight lines joining the points of intersection and the centre of the circle form a right angle.

20. If a semicircle be described on the side of a quadrant, and from any point in the quadrantal arc a radius be drawn; the part of this radius intercepted between the quadrant and semicircle, is equal to the perpendicular let fall from the same point on their common tangent.

21. If the chord of a quadrant be made the diameter of a semicircle, and from its extremities two straight lines be drawn to any point in the circumference of the semicircle; the segment of the greater line intercepted between the twc circumferences shall be equal to the less of the two lines.

22. If two circles cut each other so that the circumfer

ence of one passes through the centre of the other, and from either point of intersection a straight line be drawn cutting both circumferences; the part intersected between the two circumferences will be equal to the chord drawn from the other point of intersection to the point where it meets the inner circumference.

23. If through any point in the common chord of two circles, which intersect one another, there be drawn any two other chords, one in each circle, their four extremities shall all lie in the circumference of a circle.

24. If from the extremities of any diameter of a given circle, perpendiculars be drawn to any chord of the circle, they shall meet the chord, or the chord produced, in two points which are equidistant from the centre.

25. If two circles cut each other, and from any point in the straight line produced, which joins their intersections, two tangents be drawn, one to each circle, they shall be equal to one another.

26. If from any two points in the circumference of a circle there be drawn two straight lines to a point in a tangent to that circle, they will make the greatest angle when drawn to the point of contact.

27. If any chord in a circle be bisected by another and produced to meet the tangents drawn from the extremities of the bisecting line; the parts intercepted between the tangents and the circumferences are equal.

28. If two opposite angles of a trapezium be right angles, the angles subtended by either side at the two opposite angular points will be equal.

29. If a perpendicular be let fall from the right angle of a right-angled triangle on the hypotenuse, the rectangle contained by the hypotenuse, and either of the segments, into which it is divided by the perpendicular, is equal to the square of the side adjacent to that segment.

30. A quadrilateral figure may have a circle described about it, if the rectangle contained by the segments of the diagonal be equal.

31. If an equilateral triangle be inscribed in a circle, the square on a side thereof is equal to three times the square described upon the radius.

32. The vertical angle of any oblique-angled triangle inscribed in a circle, is greater or less than a right angle, by

the angle contained by the base and the diameter drawn from the extremity of the base.

33. If a circle be inscribed in a right-angled triangle, the difference between the two sides containing the right angle and the hypotenuse, is equal to the diameter of the circle.

34. If a semicircle be inscribed in a right-angled triangle, so as to touch the hypotenuse and perpendicular, and from the extremity of its diameter a line be drawn through the point of contact to meet the perpendicular produced; the part produced will be equal to the perpendicular.

GEOMETRICAL ANALYSIS.

In the method of analysis we assume the proposition advanced, and then proceed to trace the consequences which follow from this assumption, till we arrive at some known or admitted relation. The reverse of this process constitutes synthesis, or composition, which is the method employed in the preceding pages. In the solution of geometrical problems of more than ordinary difficulty, it is necessary that we should adopt the method of analysis, in order to discover the different steps which must be pursued in the construction. Analysis, observes an eminent geometer, presents the medium of invention; while synthesis naturally directs the course of instruction.

The following problem is given as an illustration of this method.

PROBLEM. *Given the hypotenuse of a right-angled triangle, and the sum of the base and perpendicular, to construct the triangle.*

ANALYSIS. Suppose the thing to be done. Let ABC be the triangle required, having BC equal to the given hypotenuse, the angle BAC equal to a right angle, and AB, AC together equal to the given sum of the base and perpendicular. On BA produced take AD equal to AC, and join CD.

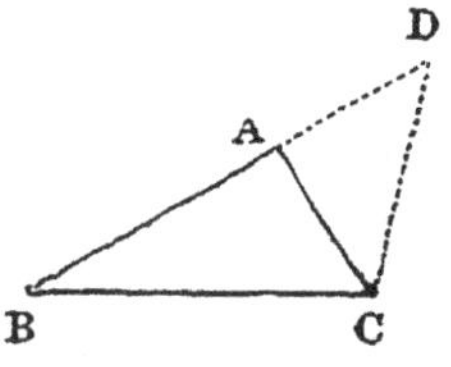

Then BD is equal to the sum of AB and AC. And since ADC is a right-angled isosceles triangle, therefore the angles ADC and ACD are each of them half a right angle.

Hence we have the following composition: —

SYNTHESIS. Make BD equal to the sum of AB and AC; from D draw (E. I. 11. and 9.) DC making the angle BDC equal to half a right angle; from B, the other extremity of BD, draw BC equal to the given hypotenuse, meeting DC in C; and from C (E. I. 12.) let fall the perpendicular CA; then ABC is the triangle required.

For the right-angled triangle ADC is evidently isosceles, and therefore AC is equal to AD, and the sum of AB and AC is equal to BD; wherefore, &c.

LONDON

PRINTED BY SPOTTISWOODE AND CO.

NEW-STREET SQUARE

Made in the USA
Monee, IL
04 November 2026

67385004R00066